Advanced Ocean Modelling

Jochen Kämpf

Advanced Ocean Modelling

Using Open-Source Software

Springer

Jochen Kämpf
Flinders University
School of the Environment
PO Box 2100 Adelaide SA 5001
Australia
jochen.kaempf@flinders.edu.au

Additional material to this book can be downloaded from http://extra.springer.com

ISBN 978-3-642-42319-2 ISBN 978-3-642-10610-1 (eBook)
DOI 10.1007/978-3-642-10610-1
Springer Heidelberg Dordrecht London New York

Softcover re-print of the Hardcover 1st edition 2010

Cover design: Bauer, Thomas

Printed on acid-free paper

Springer is part of Springer Science+Business Media (www.springer.com)

Preface

This book focuses on motions of incompressible fluids of a freely moving surface being influenced by both the Earth's rotation and density stratification. In contrast to traditional textbooks in the field of geophysical fluid dynamics, such as those by by Cushman-Roisin (1994) and Gill (1982), this book uses the method of process-oriented hydrodynamic modelling to illustrate a rich variety of fluid phenomena. To this end, the reader can adopt the model codes, found on the Springer server accompanying this book, to reproduce most graphs of this book and, even better, to create animation movies. The reader can also employ the codes as templates for own independent studies. This can be done by a lay person as a hobby activity, undergraduate or postgraduate students as part of their education, or professional scientists as part of research.

Exercises of this book are run with *open-source* software that can be freely downloaded from the Internet. This includes the FORTRAN 95 compiler "G95" used for execution of model simulations, the data visualisation program "SciLab", and "ImageMagick" for the creation of graphs and GIF animations, which can be watched with most Internet browsers.

Readers new to the subject are advised to read my book "Ocean Modelling for Beginners" (Kämpf, 2009), which gives descriptions, not replicated here, of the basics of geophysical fluid dynamics, finite-difference modelling, and the use of the above software suites. The latter book deals with so-called *layer models*, predicting the motions of multiple layers of different densities and freely moving interfaces. Such models are used in practice as forecasting tools for tides and tsunamis. The latter book also contains a detailed description of the Coriolis force. This force contributes to the *geostrophic balance* that makes the larger-scale oceanic dynamics much more structured and predictable than turbulent motions in a tea cup. The reader is advised to review the Coriolis force which plays a crucial role in many phenomena described here. This book focuses on so-called *level models* which, in contrast to layer models, are capable of simulating vertical mixing processes such as density-driven convection or breaking internal waves.

I dedicate this book to my doctor father Professor Jan O. Backhaus for his creativity and overwhelming enthusiasm which have been the prime motivation for me to pursue a career in the field of physical oceanography. I particularly remember discussions with Jan on how to include a free sea surface in a nonhydrostatic convection model, a problem that neither of us could resolve at that time. I have included such a model in this book. Other invaluable sources of motivation behind this work are the classical books of Henry Stommel, namely "An Introduction to the Coriolis Force" published in 1989 and co-authored by Dennis Moore, and "A View of the Sea", published in 1987. Similar to the approach I take here, Stommel's work underpins theory with computer programs, written in BASIC, that can be run by the reader for illustration of dynamical processes.

Adelaide, Australia, Jochen Kämpf
October 2009

Contents

Chapter 1
Introduction

Abstract This chapter reviews the Navier-Stokes equations for an incompressible fluid, summarises the basics of finite-difference modelling, and gives an overview of softwares required for the conduction of exercises.

1.1 Fundamental Physical Laws

1.1.1 Cartesian Coordinates

For convenience, locations are defined by means of the *Cartesian coordinate system* (Fig. 1.1) in which the vertical axis points upward at right angle to the undisturbed surface of a fluid at rest. Horizontal coordinates are denoted by x and y. The x-axis points to the east. The y-axis points to the north. The undisturbed surface of the fluid is defined by $z = 0$. Use of a Cartesian coordinate system implies that the true curvature of the sea surface is ignored, which is a reasonable approximation for oceanic processes on spatial scales < 500 km.

1.1.2 The Navier-Stokes Equations

The *Navier-Stokes equations* comprise several physical conservation principles; that is, conservation of momentum (Newton's laws of motion), conservation of mass (which turns into conservation of volume for an incompressible fluid), and conservation of field variables such as temperature and salinity that via the equation of state give density (which appears in the buoyancy force). In Cartesian coordinates, the momentum equations can be written as:

$$
\begin{aligned}
\frac{\partial u}{\partial t} + \mathrm{Adv}(u) - f v &= -\frac{1}{\rho_o}\frac{\partial P}{\partial x} + \mathrm{Diff}(u) \\
\frac{\partial v}{\partial t} + \mathrm{Adv}(v) + f u &= -\frac{1}{\rho_o}\frac{\partial P}{\partial y} + \mathrm{Diff}(v) \\
\frac{\partial w}{\partial t} + \mathrm{Adv}(w) &= -\frac{1}{\rho_o}\frac{\partial P}{\partial z} - \frac{(\rho - \rho_o)}{\rho_o} g + \mathrm{Diff}(w)
\end{aligned}
\tag{1.1}
$$

J. Kämpf, *Advanced Ocean Modelling*, DOI 10.1007/978-3-642-10610-1_1,

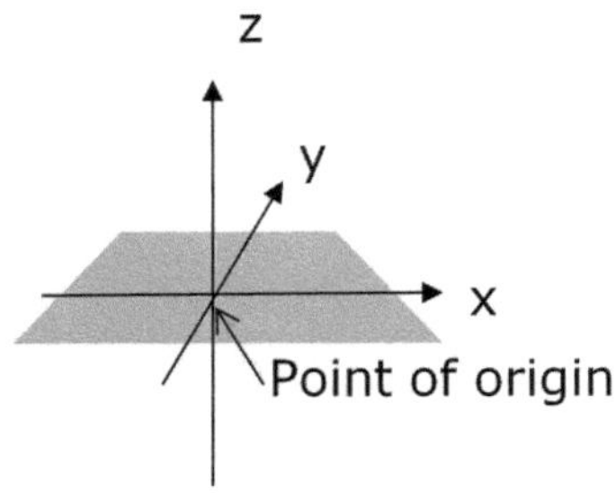

Fig. 1.1 The Cartesian coordinate system

where (x, y, z) is location in the Cartesian coordinate system, (u, v, w) is the velocity vector, t is time, f is the Coriolis parameter, P is dynamic pressure, ρ is density, mean density is ρ_o, and g is acceleration due to gravity. Density is weight of seawater per unit volume. The operator Adv() denotes the advection terms and is given by:

$$\mathrm{Adv}(\psi) = u\frac{\partial \psi}{\partial x} + v\frac{\partial \psi}{\partial y} + w\frac{\partial \psi}{\partial z}$$

where ψ is the property subject to advection. Momentum advection is also referred to as the *nonlinear terms*. Diffusion of any of the three velocity components is given by:

$$\mathrm{Diff}(\psi) = \frac{\partial}{\partial x}\left(A_h\frac{\partial \psi}{\partial x}\right) + \frac{\partial}{\partial y}\left(A_h\frac{\partial \psi}{\partial y}\right) + \frac{\partial}{\partial z}\left(A_z\frac{\partial \psi}{\partial z}\right)$$

where A_h and A_z are horizontal and vertical eddy viscosities parameterising the effects of turbulence. Dynamic pressure includes only pressure parts that have a dynamical consequence. The pressure field associated with uniform density and a plane sea surface does not contribute to the horizontal pressure-gradient force and it can therefore be subtracted from the true pressure field.

The *Boussinesq approximation*, used in the above equation, is based on the assumption that density fluctuations are small compared with mean density, which is the case for oceanic applications. To this end, density can be expressed by a constant value except when multiplied with gravity.

The essence of the momentum equations is that an imbalance of forces acting on a fluid parcel causes an acceleration or deceleration of the parcel. On the other hand, motions remain steady if the residual force vanishes, a situation referred to as *steady state*.

For an incompressible fluid, mass conservation turns in a conservation principle for volume, which can be expressed by the *continuity equation*, given by:

$$\frac{\partial u}{\partial x} + \frac{\partial v}{\partial y} + \frac{\partial w}{\partial z} = 0 \tag{1.2}$$

When ignoring slight nonlinearities and molecular (double-diffusive) effects and for an incompressible fluid, the evolution of the density field can be expressed by a *density conservation equation*, given by:

$$\frac{\partial \rho}{\partial t} + \mathrm{Adv}(\rho) = \mathrm{Diff}(\rho) \tag{1.3}$$

The diffusion operator in the latter equation is given by:

$$\mathrm{Diff}(\rho) = \frac{\partial}{\partial x}\left(K_h \frac{\partial \rho}{\partial x}\right) + \frac{\partial}{\partial y}\left(K_h \frac{\partial \rho}{\partial y}\right) + \frac{\partial}{\partial z}\left(K_z \frac{\partial \rho}{\partial z}\right)$$

where K_h and K_z are horizontal and vertical eddy diffusivities of density which can differ from eddy viscosities.

An additional equation is required for prediction of the evolution of the free sea surface, which gives the surface boundary values for dynamic pressure in the momentum equations. This equation can be derived from vertical integration of the continuity equation (Eq. 1.2) and is given by:

$$\frac{\partial \eta}{\partial t} = -\frac{\partial (h \langle u \rangle)}{\partial x} - \frac{\partial (h \langle v \rangle)}{\partial y} \tag{1.4}$$

where h is total fluid depth, and $\langle u \rangle$ and $\langle v \rangle$ are depth-averaged values of horizontal velocity components.

Depth-constant horizontal flow components are referred to as *barotropic flow*, whereas the superposed depth-variable component is called *baroclinic flow*. For purely hydrostatic dynamics, a slanted sea surface is the principal agent of barotropic flow. In contrast, baroclinic flow is created by lateral density gradients and/or frictional effects.

1.1.3 Boundary Fluxes

In addition to initial conditions, the Navier-Stokes equations require information of boundary fluxes of variables. This includes tangential frictional stresses (wind stress, bottom friction, and lateral friction), and air-sea heat and freshwater fluxes. These fluxes will be detailed in due course of this book.

1.1.4 The Hydrostatic Approximation

For processes of a horizontal length scale large compared with the vertical length scale, the momentum equation for vertical velocity w turns into the *hydrostatic approximation*, given by:

$$0 = -\frac{\partial P}{\partial z} - \rho g \tag{1.5}$$

The hydrostatic approximation forms the basis of the so-called shallow-water model and it is employed in so-called hydrostatic level models, being frequently applied by fluid modellers. In contrast, models developed in this volume are based on the full equations (Eq. 1.1) to enable the predictions of both hydrostatic and non-hydrostatic processes, a method called *nonhydrostatic modelling*. Nonhydrostatic processes include those in which horizontal and vertical scales are of similar order of magnitude.

1.1.5 The Stability Frequency

The degree of density stratification in the ocean can be characterised by the so-called *Brunt-Väisälä frequency* N, defined by:

$$N^2 = -\frac{g}{\rho_o}\frac{\partial \rho}{\partial z} \tag{1.6}$$

where ρ_o is mean density. This frequency, which appears as a characteristic scale for many stratified processes, is referred to as *stability frequency* in the following.

1.2 Numerical Methods

1.2.1 Finite Differences

Models developed in this book are based on finite-difference versions of the Navier-Stokes equations. The basis is that dynamic variables (velocity components, sea level, density, and dynamic pressure) are calculated at certain discrete grid points in space. This requires an accurate representation of both gradients (the first derivative) and curvature (the second derivative) of the spatial distribution of a variable. These derivatives can be approximated from *Taylor series*. The first spatial derivative of a variable f, for instance, with respect to a space coordinate x can be approximated in three different ways:

- $\partial f/\partial x \approx (f_{k+1} - f_k)/\Delta x$ (forward difference)
- $\partial f/\partial x \approx (f_k - f_{k-1})/\Delta x$ (backward difference)
- $\partial f/\partial x \approx (f_{k+1} - f_{k-1})/(2\Delta x)$ (centred difference)

where Δx is the spacing between adjacent grid points, called *grid spacing*, and the index k points to a certain grid cell along the x-axis. A finite-difference representation of the second spatial derivative of a function f is:

$$\frac{\partial^2 f}{\partial x^2} \approx \frac{f(x+\Delta x) - 2f(x) + f(x-\Delta x)}{(\Delta x)^2} = \frac{f_{k+1} - 2f_k + f_{k-1}}{(\Delta x)^2}$$

A certain *truncation error* is made when using these approximations. Higher-order finite-difference schemes, not detailed in this book, can be used to reduce the truncation error.

1.2.2 Requirements for a Finite-Difference Model

A finite-difference model needs to satisfy the four requirements:

- Consistency
- Accuracy
- Numerical stability
- Efficiency

The first requirement is that the finite-difference equations have to be consistent with the differential equations describing a physical process. The second requirement is that the model prediction should be as accurate as possible; that is, truncation errors and round-off errors should be kept as small as possible. The third requirement is that the prediction has to be numerically stable. Certain *stability criteria* need to be satisfied in order to achieve this. The forth requirement is that a model simulation should be as efficient as possible in terms of total simulation time and the amount of data produced.

1.3 Modelling with FORTRAN 95

1.3.1 Writing and Compiling Codes

FORTRAN 95 is used as programming language to calculate the evolution of a dynamical process being described by a set of finite-difference equations. FORTRAN codes are written as text documents saved with the file extension ".f95" for later identification. Using the open-source "G95" FORTRAN compiler, these files are converted into an executable file by entering:

g95 file.f95

in the Command Prompt window. On Microsoft Windows operation systems, the Command Prompt window is found under Start/All Programs/Accessories. The G95 FORTRAN compiler is available for many different computer platforms and can be downloaded from:

http://www.g95.org

1.3.2 Modular Source Codes

FORTRAN source codes of this book consist of several text files, a main program plus separate so-called *modules*. Most exercises of this book use two separate module files. One module contains hereby global declarations, the other one all subroutines and functions. Figure 1.2 displays these components and how they are linked with each other.

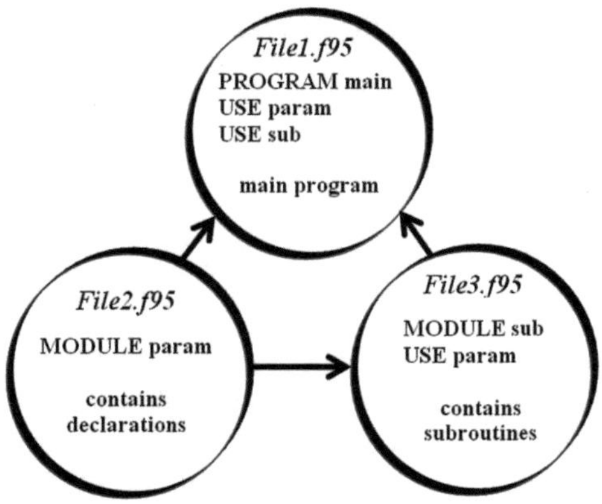

Fig. 1.2 Code structure and associated file names

The compiling of a multiple file FORTRAN code consists of two steps. The modules are first compiled with:

```
g95 -c file2.f95 file3.f95
```

Then, the module files can be linked with the main code via:

```
g95 -o run.exe file1.f95 file2.o file3.o
```

where the “-o” allows for specification of a name of the executable program, called “run.exe” in this example. The code can then be executed by entering “run.exe” in the Command Prompt window. The above example is for Microsoft Windows operation systems. Executable files carry a different file extension for other operation systems.

1.4 Visualisation with SciLab

1.4.1 Writing SciLab Scripts

The open-source SciLab software suite is used for visualisation of results. SciLab is available from:

http://www.scilab.org

SciLab scripts are formulated as text documents with the file extension ".sce". SciLab is a windows-based application software. It has a main control window, a script-based editor window, and a graphics display window. For more information on the use of SciLab, the reader should study the tutorial found on the Springer server. The author has used Version 5.1.1.

1.4.2 GIF Animations

A numbered sequence of GIF files is produced with addition of the following lines at the end of a SciLab animation loop:

```
if n < 10 then
  xs2gif(0,'Frame100'+string(n)+'.gif')
else
  if n < 100 then
    xs2gif(0,'Frame10'+string(n)+'.gif')
  else
    xs2gif(0,'Frame1'+string(n)+'.gif')
  end
end
```

In this example, "0" refers to the index of the graphics window, the index n is the loop counter used in the animation loop. This produces the sequence of GIF files: Frame1001.gif, Frame1002.gif, Frame1003.gif, etc. The open-source software ImageMagick can be downloaded from:

http://www.imagemagick.org/

This software can be used to convert these GIF frame files into a single GIF animation, that can be viewed in many Internet browsers. Conversion into an animated GIF file, called "Animation.gif", for example, is achieved by entering in the Command Prompt window:

convert -delay 10 Frame.gif Animation.gif*

where the delay option specifies the time lapse between frames in milliseconds. After creation of an animated GIF file, the reader should delete individual frame files to free up some space on the computer by entering:

del Frame.gif*

1.5 Organisation of Work

The author recommends to keep all FORTRAN, SciLab and data files belonging to an exercise in a single folder. Opening SciLab from this folder will open the SciLab program in the right directory finding the relevant data input files, previously created by a FORTRAN code. This "single-folder approach" helped the author to save valuable amounts of time that otherwise would have been wasted on endless file searches.

1.6 Download of Computer Codes

FORTRAN simulation codes and SciLab animation scripts for exercises of this book can be downloaded from:

http://extra.springer.com

A file called "Info.txt" is provided for each exercise giving further instructions. Additional codes referred to in the text are found in the folder "Miscellaneous" on the above ftp site. Files can be downloaded with right-mouse click and "Save Target as...". The author made these codes only available for "lazy" readers. Best learning outcomes are achieved when creating own codes from scratch.

Chapter 2
1D Models of Ekman Layers

Abstract This chapter introduces the reader to one-dimensional water-column models using fixed vertical levels. Such a model is applied to study the dynamics of surface and bottom Ekman layers in the ocean.

2.1 Useful Background Knowledge

2.1.1 Inertial Oscillations

Before exploring the Ekman-layer dynamics, it is useful to revisit features inherent with inertial oscillations. Flows under the sole influence of the Coriolis force are described by the momentum equations:

$$\frac{\partial u}{\partial t} + fv = 0 \tag{2.1}$$

$$\frac{\partial v}{\partial t} - fu = 0 \tag{2.2}$$

where f is the Coriolis parameter, given by $f = 2\Omega \sin(\varphi)$, where $\Omega = 7.27 \times 10^{-5}\mathrm{s}^{-1}$ is the rotation frequency of Earth, and φ is geographical latitude in radians. For an initial flow in the x-direction of speed u_o, the solution of the latter equations is given by:

$$u(t) = +u_o \cos(ft)$$
$$v(t) = -u_o \sin(ft)$$

The resultant flow trajectories are circles of a radius of $u_o/|f|$, called *inertial radius*. The period of one complete cycle is $2\pi/f$, called *inertial period*. The inertial period is 12 hrs at the poles and infinite directly at the equator where the Coriolis force vanishes. Figure 2.1 shows flow paths associated with inertial oscillations for $u_o = 0.1$ m/s and $f = 1 \times 10^{-4}\mathrm{s}^{-1}$ with and without ambient uniform flow.

J. Kämpf, *Advanced Ocean Modelling*, DOI 10.1007/978-3-642-10610-1_2,

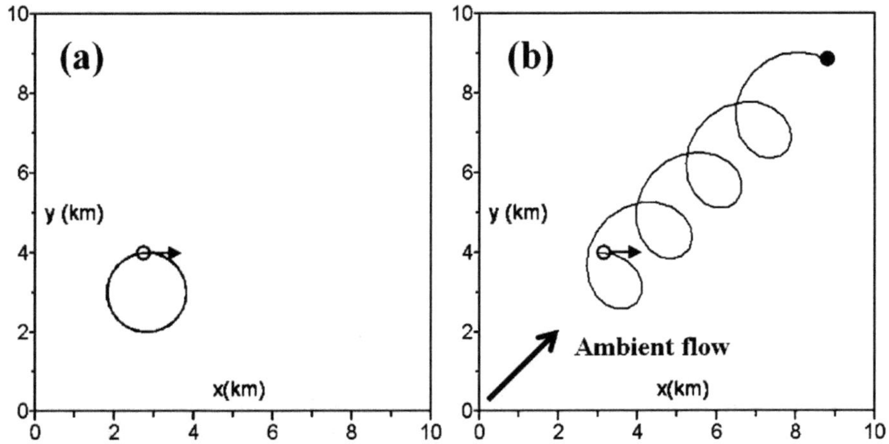

Fig. 2.1 Examples of flow paths induced by inertial oscillations (**a**) in the absence of ambient flow, and (**b**) for an ambient flow of $u_{amb} = v_{amb} = 0.02\,\text{m/s}$. The *open circle* shows the starting position of a float, the *closed circle* the end position

2.1.2 Semi-implicit Treatment of the Coriolis Force

Adequate formulation of the Coriolis force in a finite-difference model can be achieved by means of a semi-implicit approach. For the momentum equations governing inertial oscillations (Eqs. 2.1 and 2.2), this approach gives:

$$
\begin{aligned}
u^{n+1} &= u^n + 0.5\,\alpha(v^n + v^{n+1})\\
v^{n+1} &= v^n - 0.5\,\alpha(u^n + u^{n+1})
\end{aligned}
$$

where n is the current time level, $n+1$ refers to the future value (one time step Δt ahead), and $\alpha = \Delta t f$. Cross-combination of the latter equations yields the final form:

$$
\begin{aligned}
u^{n+1} &= \left[(1-\beta)u^n + \alpha v^n\right]/(1+\beta)\\
v^{n+1} &= \left[(1-\beta)v^n - \alpha u^n\right]/(1+\beta)
\end{aligned}
$$

where $\beta = 0.25\,\alpha^2$. This scheme requires numerical time steps small compared with the rotation period; that is, $|\alpha| << 1$, otherwise the period of the parcel's circular motion will differ from the true value. This semi-implicit scheme is widely used by modellers.

2.2 The Surface Ekman Layer

2.2.1 Boundary-Layer Equations

This section explores the dynamics of frictional boundary layers in the ocean, called *Ekman layers* (Ekman, 1905). For simplicity, we assume horizontal homogeneity of all variables and an ocean of uniform density, so that the Navier-Stokes equations take the reduced form:

$$\frac{\partial u}{\partial t} - f v = \frac{\partial}{\partial z}\left(A_z \frac{\partial u}{\partial z}\right) \tag{2.3}$$

$$\frac{\partial v}{\partial t} + f u = \frac{\partial}{\partial z}\left(A_z \frac{\partial v}{\partial z}\right) \tag{2.4}$$

where f is the Coriolis parameter, and the terms on the right-hand side of these *boundary-layer equations* represent vertical turbulent diffusion of momentum with A_z being vertical eddy viscosity.

Wind stress operates as a tangential frictional force at the sea surface, and the associated boundary conditions read:

$$\left(A_z \frac{\partial u}{\partial z}\right)_{z=0} = \frac{\tau_x^{\text{wind}}}{\rho_o} \text{ and } \left(A_z \frac{\partial v}{\partial z}\right)_{z=0} = \frac{\tau_y^{\text{wind}}}{\rho_o} \tag{2.5}$$

where ρ_o is surface density. The components of the wind-stress vector are given by:

$$\tau_x^{\text{wind}} = \rho_{\text{air}} C_d\, U \sqrt{U^2 + V^2} \text{ and } \tau_y^{\text{wind}} = \rho_{\text{air}} C_d\, V \sqrt{U^2 + V^2} \tag{2.6}$$

where ρ_{air} is air density, C_d is the nondimensional *wind-drag coefficient* with values in a range of $1.1 - 1.5 \times 10^{-3}$, and U and V are horizontal components of the wind vector measured at a height of 10 m above sea level. The wind stress vector field has the same direction as the wind, but its magnitude is proportional to the square of the wind speed. Hence, the stronger the wind the greater are its impacts on surface flows.

The wind stress creates tangential friction along the sea surface and, thus, transfers momentum into the ocean. On time scales of days, the resultant oceanic motion becomes influenced by the Coriolis force. In the absence of other influences, final steady state consists of a dynamical balance between the Coriolis force and the friction force. What is the structure of the resultant steady-state flow pattern?

2.2.2 Scaling: The Temporal Rossby Number

Consider an oscillatory flow of a maximum speed of U_o on a period of T. On the basis of this, the Coriolis force attains a maximum value of $f U_o$. On the other hand, the order of magnitude of the temporal derivative in Eq.(2.3) can be estimated at:

$$\frac{\partial u}{\partial t} \approx \frac{U_o}{T}$$

The ratio of this estimate with that of the Coriolis force in the boundary-layer equations is given by:

$$Ro_t = \frac{1}{f\,T} \tag{2.7}$$

and is called the *temporal Rossby number*. This comparative ratio implies that the Coriolis force can no longer be neglected in the momentum equations if $Ro_t \approx 1$, or, in other words, if the time scale of a process (establishment of a frictional boundary layer here) is of the order of the inertial period, given by $2\pi/\text{f}$. Hence, except for the equatorial regions, where the inertial period becomes long, the Coriolis force becomes important if a flow lasts longer than a few days. Considerations based on typical *scales of motion* and comparative ratios of terms in the momentum equations are called *scaling* considerations.

2.2.3 Scaling: The Ekman Number

For a steady state, the Coriolis force is balanced by the frictional force associated with vertical diffusion of momentum. The form of the boundary equations (Eqs. 2.3 and 2.4) implies that lateral velocity varies exponentially with depth. With such a velocity profile; that is,

$$u(z) = U_o \exp(z/D)$$

where U_o is the surface value and D is a depth scale, the magnitude of the frictional force is $A_z U_o/D^2$, assuming vertical eddy viscosity to be uniform. The ratio between this magnitude with that of the Coriolis force is called the *Ekman number* and is given by:

$$Ek = \frac{A_z}{D^2 f} \tag{2.8}$$

Accordingly, a steady state of the boundary-layer equations (Eqs. 2.3 and 2.4) implies that $Ek \approx 1$, which corresponds to a depth scale of:

$$D = \sqrt{\frac{A_z}{f}} \tag{2.9}$$

Consequently, surface Ekman layers in the ocean are limited in their vertical extent. Typical thicknesses are 50–150 m with increasing values toward the equator.

2.2.4 Solutions of the Boundary-Layer Equations

The analytical solutions of the boundary-layer equations are derived and discussed in many textbooks (e.g. Cushman-Roisin, 1994). These solutions are a first benchmark for verification of a one-dimensional finite-difference water-column model being constructed in the next section. This approach also gives us the opportunity to explore situations of nonuniform values of vertical eddy viscosity for which analytical solutions are more difficult to derive.

2.2.5 Finite-Difference Equations

The water column is represented by grid cells stacked on top of each other with a vertical grid spacing of Δz and a grid index i with $i = 1$ pointing to the surface cell and $i = nz$ to the bottom cell (Fig. 2.2). In this one-dimensional application, values of velocity components and eddy viscosity are calculated at the same grid point. The uppermost grid point is located at a distance of $0.5\Delta z$ below the sea surface.

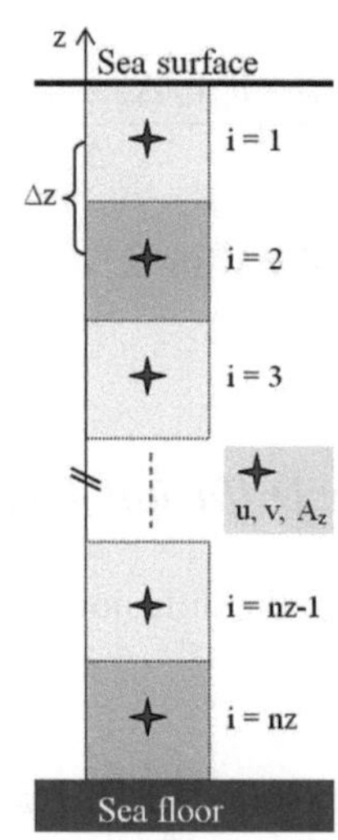

Fig. 2.2 Grid index, grid points and grid spacing for Ekman-layer modelling

Use of the semi-implicit approach for the Coriolis force leads to the finite-difference equations:

$$u_i^{n+1} = u_i^n + 0.5\,\alpha(v_i^n + v_i^{n+1}) + \mathrm{Diff}_u \tag{2.10}$$
$$v_i^{n+1} = v_i^n - 0.5\,\alpha(u_i^n + u_i^{n+1}) + \mathrm{Diff}_v \tag{2.11}$$

where n is the current time level, $n + 1$ refers to the future value (one time step Δt ahead), $\alpha = \Delta t f$, and Diff_u and Diff_v are the diffusion terms. Cross-combination of the latter equations gives:

$$u_i^{n+1} = \left[(1-\beta)u_i^n + \alpha v_i^n + 0.5\alpha \mathrm{Diff}_v + \mathrm{Diff}_u\right]/(1+\beta) \tag{2.12}$$
$$v_i^{n+1} = \left[(1-\beta)v_i^n - \alpha u_i^n - 0.5\alpha \mathrm{Diff}_u + \mathrm{Diff}_v\right]/(1+\beta) \tag{2.13}$$

with $\beta = 0.25\,\alpha^2$. Accurate representation of the Coriolis force requires $\Delta t << 1/\,|f|$.

2.2.6 *Formulation of Diffusion Terms*

In finite-difference form, the diffusion terms can be written as:

$$\mathrm{Diff}_u = \Delta t\,\frac{A_z^+\left(u_{i-1}^n - u_i^n\right)/\Delta z - A_z^-\left(u_i^n - u_{i+1}^n\right)/\Delta z}{\Delta z} \tag{2.14}$$

$$\mathrm{Diff}_v = \Delta t\,\frac{A_z^+\left(v_{i-1}^n - v_i^n\right)/\Delta z - A_z^-\left(v_i^n - v_{i+1}^n\right)/\Delta z}{\Delta z} \tag{2.15}$$

where A_z is vertical eddy viscosity with $A_z^+ = 0.5(A_{z,i-1} + A_{z,i})$ and $A_z^- = 0.5(A_{z,i} + A_{z,i+1})$.

2.2.7 *Stability Criterion for Diffusion Terms*

The above finite-difference form of the diffusion terms is associated with the stability criterion:

$$\Delta t \le \frac{(\Delta z)^2}{A_{z,\max}} \tag{2.16}$$

where $A_{z,\max}$ is the maximum value that vertical eddy viscosity attains during a simulation. The time step chosen has to satisfy Eq. 2.16, otherwise the prediction becomes numerically unstable causing the computer code to crash.

2.3 Exercise 1: The Surface Ekman Layer

2.3.1 Task Description

We consider a water column, 500 m in depth, represented by an equidistant vertical grid spacing of 1 m. The Coriolis parameter is chosen as $f = 1 \times 10^{-4}\text{s}^{-1}$ corresponding to mid-latitudes in the Northern Hemisphere. The water column is initially at rest. The model is forced via prescription of a southerly wind of a wind stress of $\tau_y = 0.5$ Pa in magnitude.

The total simulation time is 5 days. To avoid the appearance of strong inertial oscillations, the wind stress is linearly adjusted from zero to its final value over the first 2 days of simulation. The time step is set to $\Delta t = 5$ s.

The surface wind stress enters the finite-difference equations implicitly via the boundary values u_0^n and v_0^n. Using Eq. (2.5), these values are calculated from:

$$u_0^n = u_1^n + \frac{\tau_x^{\text{wind}}}{\rho_o\, A_z^+} \Delta z \tag{2.17}$$

$$v_0^n = v_1^n + \frac{\tau_y^{\text{wind}}}{\rho_o\, A_z^+} \Delta z \tag{2.18}$$

where $A_z^+ = 0.5(A_{z,0} + A_{z,1})$ with $A_{z,0}$ representing vertical eddy viscosity near the sea surface. The following three different eddy-viscosity scenarios are considered in this exercise:

1. Eddy viscosity is uniform with a constant value of $A_z = 5 \times 10^{-2}\,\text{m}^2\,\text{s}^{-1}$;
2. Same as before, but with a local minimum of $A_z = 4 \times 10^{-3}\,\text{m}^2\,\text{s}^{-1}$ around a depth of 20 m mimicking a reduction of turbulence levels by an assumed strong local density stratification;
3. Eddy viscosity is calculated from Prandtl's mixing-length approach (Prandtl, 1925) according to:

$$A_z = L^2\sqrt{(\partial u/\partial z)^2 + (\partial v/\partial z)^2}$$

where, for simplicity, the mixing length is set to a constant value of $L = 2$ m.

Only results of the first scenario are presented here. The other scenarios are included as options in the FORTRAN 95 code and remain for the reader to be tested. The resultant steady-state flow pattern is visualised via displacements of neutrally buoyant floats. To this end, a prediction scheme for neutrally buoyant floats is added to the code. Initially, floats form a vertical line and lateral displacements are predicted with:

$$X^{n+1} = X^n + \Delta t\ u$$
$$Y^{n+1} = Y^n + \Delta t\ v$$

where (X, Y) is the location of a float, and (u, v) is horizontal velocity at the depth horizon of this float.

2.3.2 Results

The wind-stress forcing imposed creates a lateral current in the ocean being strongest at the surface and decreasing rapidly with depth (Fig. 2.3). The surface current is directed 45° to the right with respect to the wind direction in the Northern Hemisphere. The flow direction turns clockwise with increasing distance from the sea surface. The final shape is called the *Ekman spiral*. The *Ekman-layer depth* can be defined as the depth at which the speed of the Ekman flow has decreased to $\exp(-\pi) \approx 0.04$ (4%) of its surface value. According to theory, this takes place at a depth of:

$$\delta_E = \pi \sqrt{\frac{2A_z}{|f|}} \tag{2.19}$$

yielding 100 m for the setting of this exercises. This is in excellent agreement with the simulation result. Note that some textbooks define the Ekman-layer depth without the π multiplier.

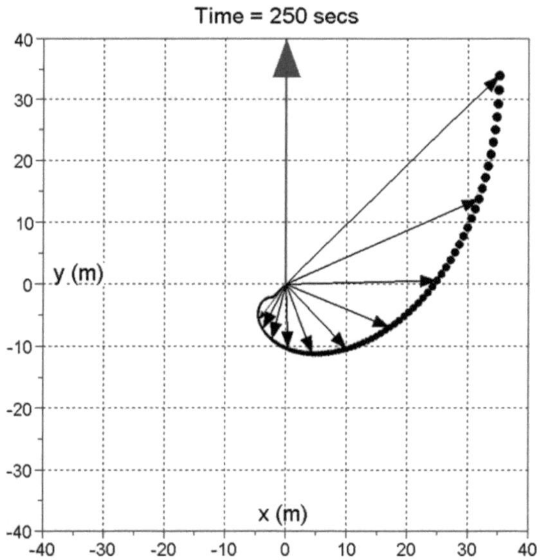

Fig. 2.3 Exercise 1. Structure of the surface Ekman layer (Northern Hemisphere). *Small arrows* indicate lateral float displacements shown from the surface to a depth of 100 m at steps of 10 m. The *thick arrow* indicates the wind direction

2.3.3 Explanation of the Ekman-Layer Structure

Imagine that the ocean consists of multiple thin layers piled up on top of each other (Fig. 2.4a). Each layer is forced by the overlying layer by a tangential stress and, itself, is subject to friction with the layer underneath. Friction operates opposite to the drift direction of a layer. The Coriolis force acts perpendicular to this direction, to the right in the Northern Hemisphere and to the left in the Southern Hemisphere. A balance of forces (Fig. 2.4b) implies now that the direction of movement of a layer is turned by a certain fraction clockwise (Northern Hemisphere) with respect to the overlying layer.

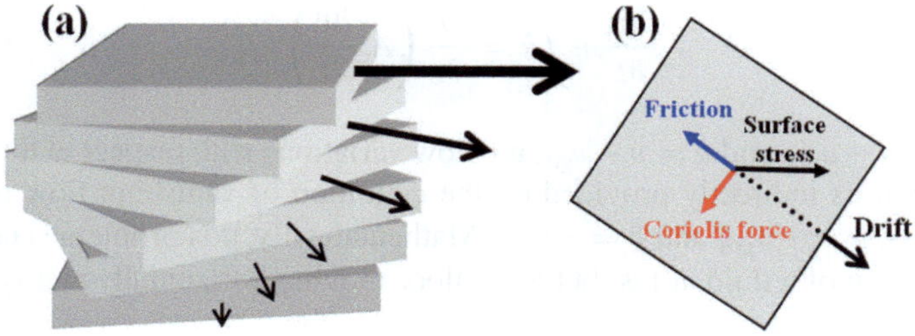

Fig. 2.4 Illustration of the structure of the surface Ekman layer in the Northern Hemisphere

The friction force at the bottom of a layer acts as a surface stress (in the opposite direction) for the next deeper layer. Since this surface stress forms the longest side (hypotenuse) of the right-angled force triangle, this tangential stress has to decrease from one layer to the next deeper one such that the drift speed decreases with depth. Consequently, the resultant layer motions make up an Ekman spiral.

2.3.4 Additional Exercises for the Reader

Repeat the simulation for the other eddy-viscosity scenarios outlined above. Calculate depth-averaged values of the components of horizontal velocity and produce data outputs on hourly intervals. Theory suggest that the depth-averaged flow in the Ekman layer is at right angle with respect to the wind direction, to the right in the Northern Hemisphere and to the left in the Southern Hemisphere (e.g. Pond and Pickard, 1983). Does the model yield the same result? If not, explore the time-average values of the results and consider what you have learned about inertial oscillations.

2.4 The Bottom Ekman Layer

2.4.1 Boundary-Layer Equations

A bottom Ekman layer develops owing to frictional effects experienced by a flow in contact with the sea floor. For simplicity, this ambient (geostrophic) flow of vector components u_{geo} and v_{geo} is assumed to be horizontally uniform and the sea floor is assumed to be plane. With horizontal homogeneity of all variables, the boundary-layer equations can then be formulated as:

$$\frac{\partial \hat{u}}{\partial t} - f\hat{v} = \frac{\partial}{\partial z}\left(A_z \frac{\partial \hat{u}}{\partial z}\right) \tag{2.20}$$

$$\frac{\partial \hat{v}}{\partial t} + f\hat{u} = \frac{\partial}{\partial z}\left(A_z \frac{\partial \hat{v}}{\partial z}\right) \tag{2.21}$$

where $\hat{u} = u - u_{geo}$ and $\hat{v} = v - v_{geo}$ are flow variations with respect to the ambient flow. Forcing is indirectly provided by the condition of vanishing flow at the sea floor; that is, $\hat{u} = -u_{geo}$ and $\hat{v} = -v_{geo}$. Mathematically, this problem is equivalent to the situation of a fluid at rest but a sea floor moving horizontally at a velocity of $(-u_{geo}, -v_{geo})$.

2.5 Exercise 2: The Bottom Ekman Layer

2.5.1 Task Description

Take the same grid configuration and initial conditions as in Exercise 1. Forcing is provided by prescription of an ambient uniform geostrophic flow of a speed of $v_{geo} = 0.1$ m/s directed to the north. For simplicity, uniform vertical eddy viscosity is used with a value of $A_z = 5 \times 10^{-3}\,\mathrm{m^2\,s^{-1}}$. The time step is set to $\Delta t = 5$ s. The total simulation time is 5 days with data output of the flow field at the end of the simulation. Note that implementation of a bottom-friction law is not necessary here, since bottom friction arises implicitly via prescription of a near-bottom value of eddy viscosity in conjunction with the assumption of vanishing flow at the sea floor.

2.5.2 Results

Like in the surface Ekman layer, the Coriolis force produces an Ekman spiral near the sea floor (Fig. 2.5). In the bottom layer, however, the flow turns toward the left with respect to the direction of the ambient flow as we move closer to the sea floor. Here, it is the sea floor that imposes a frictional stress on the fluid. In this sense, the dynamics that make up the bottom Ekman layer is very similar to features of the

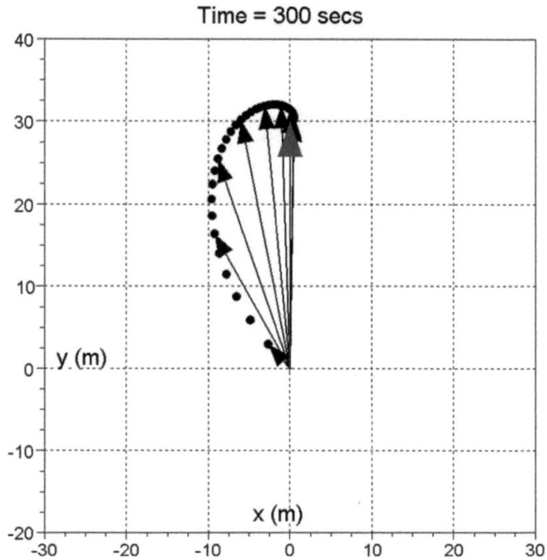

Fig. 2.5 Exercise 2. Structure of the bottom Ekman layer (Northern Hemisphere) *Small arrows* indicate lateral float displacements shown from the sea floor upward at steps of 5 m. The *thick arrow* indicates the displacement of a fluid parcel carried by ambient flow above the bottom Ekman layer

surface Ekman layer when turning the water column upside down and when imagining a sea floor moving at a speed of 0.1 m/s along the "surface" of a fluid being initially at rest. As for the surface Ekman layer, (2.19) determines the thickness of the Ekman layer. Bottom Ekman layers in the ocean attain thicknesses of 5–50 m. In this exercise, we yield a thickness of around 20 m.

2.5.3 Additional Exercises for the Reader

Consider a situation of a southerly wind of a wind-stress magnitude of $\tau_y = 0.5$ Pa in magnitude in conjunction with a northward ambient geostrophic flow of $v_{geo} = 0.1$ m/s in speed. Conduct a sequence of experiments with total water depth h varying between 20 and 200 m. For which value of h do the surface and bottom Ekman layers appear as separate features without overlapping? The reader is also encouraged to simulate Ekman-layer dynamics for the Southern-Hemisphere situation.

Chapter 3
Basics of Nonhydrostatic Modelling

Abstract This chapter introduces the reader to nonhydrostatic finite-difference solvers of the Navier-Stokes equations. For a start, the ocean is treated as as vertical slice. Flow and gradients of variables normal to this plane are assumed to vanish, and the Coriolis force is ignored. Exercises address deep-water (short) surface gravity waves, bottom-attached density-driven currents, internal waves, instabilities of vertical shear flows, lee waves, double-diffusive instability, double-diffusive layering and free convection.

3.1 Level Models

The most common types of vertical level models used in the field of oceanography are *z-coordinate models* and *σ-coordinate models* (Fig. 3.1). *Z*-coordinate models are based on fixed vertical levels of vertical grid spacing Δz. A shortcoming of z-coordinate models is a step structure of bottom topography which can lead to certain problems. In contrast to this, coordinate surfaces in σ-coordinate models follow the sea floor. The σ-coordinate is defined by:

$$\sigma = \frac{h_o + z}{h_o + \eta}$$

where z is the Cartesian vertical coordinate, h_o is undisturbed water depth, and η is sea-surface elevation. Accordingly, σ varies from zero at the sea floor to unity at the sea surface. Note that σ-coordinates respond to fluctuations of the sea level. Hydrodynamic models based on σ-coordinates are suitable for coastal applications, but steep bottom slopes can cause substantial truncation errors. Models based on σ-coordinates are not considered in this book.

J. Kämpf, *Advanced Ocean Modelling*, DOI 10.1007/978-3-642-10610-1_3,

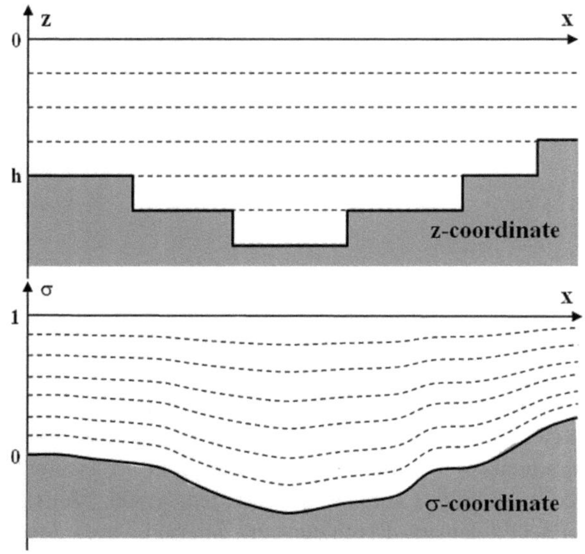

Fig. 3.1 Illustration of vertical levels in z-coordinate models and in σ-coordinate models

3.2 2D Vertical-Slice Modelling

3.2.1 Configuration

We consider a vertical ocean slice with the z-axis pointing upward and the x-axis being the horizontal coordinate (Fig. 3.2). The free sea surface is located at $z = \eta$. The sea floor can be found at $z = -h_o$. The assumption of this chapter is absence of both motion and gradients of variables perpendicular to this slice. This implies neglect of the Coriolis force. A modification to the latter are so-called *2.5-dimensional ocean-slice models* which allow for flow normal to the slice and inclusion of the Coriolis force, but this flow is also assumed to have vanishing gradients in the y-direction. Chapter 4 presents applications of such models. The momentum equations for the two-dimensional ocean slice are given by:

$$\begin{aligned}
\frac{\partial u}{\partial t} + \mathrm{Adv}(u) &= -\frac{1}{\rho_o}\frac{\partial P}{\partial x} + \frac{\partial}{\partial x}\left(A_h\frac{\partial u}{\partial x}\right) + \frac{\partial}{\partial z}\left(A_z\frac{\partial u}{\partial z}\right) \\
\frac{\partial w}{\partial t} + \mathrm{Adv}(w) &= -\frac{1}{\rho_o}\frac{\partial P}{\partial z} - \frac{\rho'}{\rho_o}g + \frac{\partial}{\partial x}\left(A_h\frac{\partial w}{\partial x}\right) + \frac{\partial}{\partial z}\left(A_z\frac{\partial w}{\partial z}\right) \qquad (3.1) \\
\frac{\partial \rho'}{\partial t} + \mathrm{Adv}(\rho') &= \frac{\partial}{\partial x}\left(K_h\frac{\partial \rho'}{\partial x}\right) + \frac{\partial}{\partial z}\left(K_z\frac{\partial \rho'}{\partial z}\right)
\end{aligned}$$

The advection operator is given by:

$$\mathrm{Adv}(\psi) = u\frac{\partial \psi}{\partial x} + w\frac{\partial \psi}{\partial z}$$

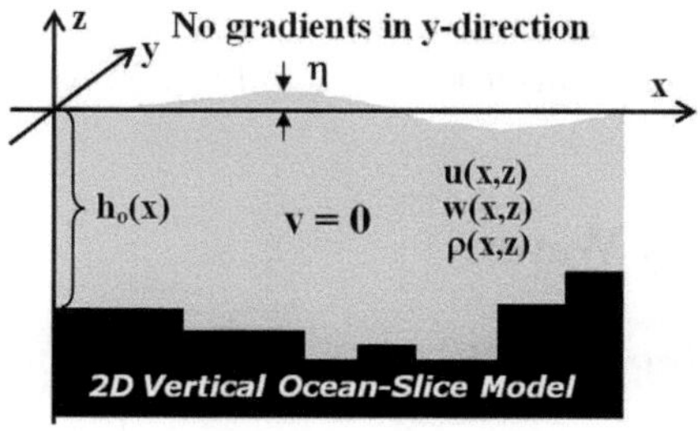

Fig. 3.2 The two-dimensional vertical ocean slice

The continuity equation (Eq. 1.2) can be written as:

$$\frac{\partial u}{\partial x} + \frac{\partial w}{\partial z} = 0 \tag{3.2}$$

3.2.2 The Arakawa C-Grid

Figure 3.3 shows the configuration of the Arakawa C-grid (Arakawa and Lamb, 1977) applied to the vertical ocean slice. This grid is the basis for all exercises of this book. Vertical location is defined by the level index i and vertical grid spacing Δz. The uppermost grid cell carries the index $i = 1$, whereas $i = nz$ points to the bottom layer. Grid points for pressure and other scalars (i.e. density) are centred between u- and w-velocity grid points. The undisturbed sea surface is aligned with vertical velocity grid points of uppermost grid cells.

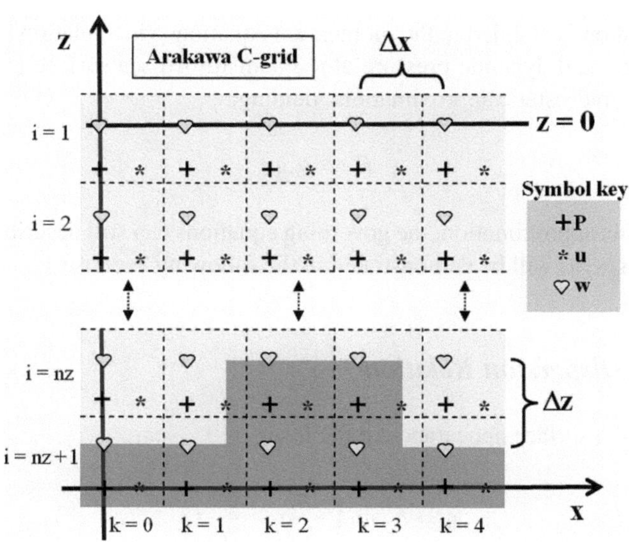

Fig. 3.3 Arakawa C-grid for a vertical ocean slice

Bottom topography is implicitly defined by setting velocity components normal to solid boundaries to zero. This leads to a step-like representation of the sea floor.

3.3 Surface Gravity Waves

3.3.1 The Governing Equations

Plane waves are waves of unidirectional propagation. Crests and troughs are oriented perpendicular to this direction. Plane surface gravity waves propagating in the x-direction in a fluid of uniform density can be described by the simplified Navier-Stokes equations:

$$\begin{aligned} \frac{\partial u}{\partial t} &= -\frac{1}{\rho_o}\frac{\partial P}{\partial x} \\ \frac{\partial w}{\partial t} &= -\frac{1}{\rho_o}\frac{\partial P}{\partial z} \\ \frac{\partial u}{\partial x} + \frac{\partial w}{\partial z} &= 0 \end{aligned} \tag{3.3}$$

where u is horizontal velocity, w is vertical velocity, ρ_o is a constant reference density, and P is dynamic pressure. For simplicity, nonlinear terms and frictional effects have been neglected here to first-order approximation.

The evolution of sea level is described by the volume-conservation equation:

$$\frac{\partial \eta}{\partial t} = -\frac{\partial (h\,\langle u \rangle)}{\partial x} \tag{3.4}$$

This equation is coupled to the momentum equations via a relation between sea-level elevation and dynamic pressure at the undisturbed sea surface ($z = 0$). Here, we apply the hydrostatic approximation, yielding:

$$P_s = \rho_o g \eta \tag{3.5}$$

Despite this approximation, the governing equations can still describe nonhydrostatic processes, as will be demonstrated in the following exercise.

3.3.2 The Dispersion Relation

For a wave of a surface appearance of the form:

$$\eta = \eta_o \sin\left(\frac{2\pi}{\lambda}x - \frac{2\pi}{T}t\right) \tag{3.6}$$

where η_o is wave amplitude, λ is wavelength and T is wave period, the solution of Eqs. (3.3, 3.4, and 3.5) are surface gravity waves that for an ocean of uniform depth h obey the dispersion relation (e.g. Pond and Pickard, 1983):

$$c = \frac{\lambda}{T} = \sqrt{\frac{g\lambda}{2\pi}\tanh\left(2\pi\frac{h}{\lambda}\right)} \tag{3.7}$$

where c is the phase speed of the wave. Figure 3.4 displays the phase speed of surface gravity waves as a function of total water depth for selected wavelengths. The dispersion relation includes two different breeds of surface gravity waves that exist in the ocean. The ratio between wavelength and total water depth determines which breed dominates. The first breed are *shallow-water waves* (or *long waves*) which can be characterised by $\lambda > 20h$. These waves are almost barotropic; that is, horizontal flow under a wave is uniform with depth, and attain a phase speed of:

$$c_{\text{long}} = \sqrt{gh} \tag{3.8}$$

Shallow-water waves are almost hydrostatic, which implies $\partial P/\partial z = 0$ in Eq. 3.3. Accordingly, horizontal pressure gradients imposed by a tilted sea surface do not vary with depth for such waves. This *hydrostatic assumption* is the basis of the shallow-water layer models employed in Kämpf (2009).

Nonhydrostatic effects lead to a second breed of gravity waves, called *deep-water waves* or *short waves*. Short waves can be classified by $\lambda < 2h$ and attain a phase speed of:

$$c_{\text{short}} = \sqrt{g\frac{\lambda}{2\pi}} \tag{3.9}$$

In contrast to shallow-water waves, deep-water waves are *dispersive*; that is, waves of greater wavelength propagate at a faster speed. An example of deep-water waves are wind-generated waves in the open ocean.

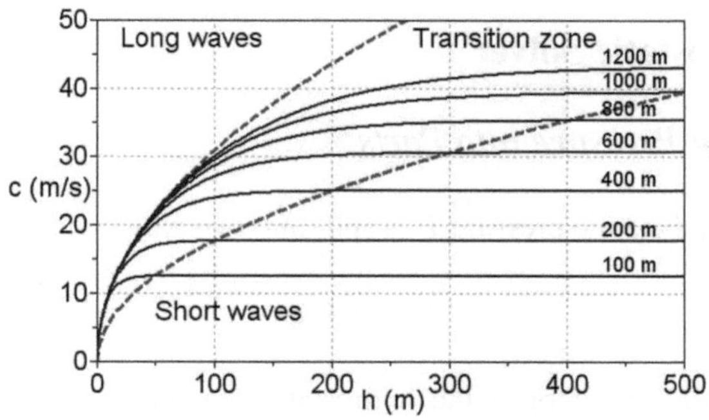

Fig. 3.4 Phase speed of surface gravity waves (*solid lines*) versus total water depth h for various wavelengths λ. *Dashed lines* show values for $\lambda = 20h$ and $\lambda = 2h$

The folder "Miscellaneous/Dispersion Relation Calculator" on the book's ftp site contains a SciLab script with which the reader can compute either phase speed or period of surface gravity waves from user-specified values of total water depth and wavelength.

3.3.3 Orbital Motions of Water Particles and Wave Pressure

Orbital motions of water parcels under a pure sine wave are described by the equations (e.g. Pond and Pickard, 1983):

$$u = 2\pi \frac{\eta_o}{T} \exp\left(-2\pi \frac{z^*}{\lambda}\right) \sin\left(2\pi \frac{x}{\lambda} - 2\pi \frac{t}{T}\right) \tag{3.10}$$

$$w = 2\pi \frac{\eta_o}{T} \exp\left(-2\pi \frac{z^*}{\lambda}\right) \cos\left(2\pi \frac{x}{\lambda} - 2\pi \frac{t}{T}\right) \tag{3.11}$$

where z^* is (positive) distance from the sea surface. The pressure field that drives these motions is given by:

$$P = \rho_o g \, \eta_o \exp\left(-2\pi \frac{z^*}{\lambda}\right) \sin\left(2\pi \frac{x}{\lambda} - 2\pi \frac{t}{T}\right) \tag{3.12}$$

which is consistent with Eq. (3.5).

For deep-water waves, water particles move in circular orbits with a radius of orbits decreasing rapidly (exponentially) with depth. At a depth $z^* = \lambda$, for instance, the orbit's radius is only 0.2% of that at the surface. This implies that such waves attain vanishingly small orbital speeds near the seafloor.

Orbital motion in shallow-water waves are elliptical near the sea surface and horizontal (simply back and forth) at the sea bottom. Shallow-water waves, if energetic enough, are capable of eroding sediment from the sea floor.

3.4 Nonhydrostatic Solver

3.4.1 Splitting Pressure into Parts

For convenience, dynamic pressure P is split into two parts:

$$P = p + q \tag{3.13}$$

where (lower-case) p refers to the hydrostatic pressure field with reference to an undisturbed (horizontal) sea level, and q includes pressure components both imposed by a tilted sea surface and arising from nonhydrostatic effects. For a constant-density ocean, the latter equation turns into $P = q$.

3.4.2 Starting as Simple as Possible

The focus is now placed on Eqs. (3.3, 3.4, and 3.5) that govern the dynamics of surface gravity waves. Eqs. (3.4) and (3.5) can be combined into a prognostic equation for the surface value of dynamic pressure q_s:

$$\frac{\partial q_s}{\partial t} = -\rho_o g \frac{\partial (h \langle u \rangle)}{\partial x} \tag{3.14}$$

Accordingly, the complete set of equations governing the dynamics of linear surface gravity waves in an ocean uniform in density is given by:

$$\frac{\partial u}{\partial t} = -\frac{1}{\rho_o}\frac{\partial q}{\partial x} \tag{3.15}$$

$$\frac{\partial w}{\partial t} = -\frac{1}{\rho_o}\frac{\partial q}{\partial z} \tag{3.16}$$

$$\frac{\partial u}{\partial x} + \frac{\partial w}{\partial z} = 0 \tag{3.17}$$

$$\frac{\partial q_s}{\partial t} = -\rho_o g \frac{\partial (h \langle u \rangle)}{\partial x} \tag{3.18}$$

The result are four coupled partial differential equations with four unknowns. The equations describe the dynamics of both short and long surface gravity waves. Unfortunately, these equations cannot be solved in a straight-forward explicit manner, because dynamic pressure appears implicitly on the right-hand side of the momentum equations.

3.4.3 Finite-Difference Scheme

The pressure part q is decomposed into contributions from the current time level (n) plus pressure corrections considering the next time level $(n+1)$. This can be written as:

$$q \Rightarrow q^n + \Delta q^{n+1} \tag{3.19}$$

Accordingly, the numerical solver of Eqs. (3.15, 3.16, 3.17, and 3.18) can be formulated in two separate steps. In the first step, a first-guess velocity is calculated explicitly from values known at time level n. In the second step, the pressure correction Δq is calculated implicitly from the requirement that the new velocity field has to be free of divergence according to Eq. (3.17). The first guess of velocity is hereby calculated from the time-forward scheme:

$$u^*_{i,k} = u^n_{i,k} - \frac{\Delta t}{\rho_o \Delta x}(q^n_{i,k+1} - q^n_{i,k}) \tag{3.20}$$

$$w^*_{i,k} = w^n_{i,k} - \frac{\Delta t}{\rho_o \Delta z}(q^n_{i-1,k} - q^n_{i,k}) \tag{3.21}$$

where i and k are the cell references for the Arakawa C-grid (see Fig. 3.3). Consequently, the finite-difference form of the momentum equations can be written as:

$$u^{n+1}_{i,k} = u^*_{i,k} - \frac{\Delta t}{\rho_o \Delta x}\left(\Delta q^{n+1}_{i,k+1} - \Delta q^{n+1}_{i,k}\right) \tag{3.22}$$

$$w^{n+1}_{i,k} = w^*_{i,k} - \frac{\Delta t}{\rho_o \Delta z}\left(\Delta q^{n+1}_{i-1,k} - \Delta q^{n+1}_{i,k}\right) \tag{3.23}$$

Insertion of the latter equations in the continuity equation (Eq. 3.17) and multiplication with the product $\Delta z \Delta x$ gives:

$$a_e \Delta q^{n+1}_{i,k+1} + a_w \Delta q^{n+1}_{i,k-1} + a_t \Delta q^{n+1}_{i-1,k} + a_b \Delta q^{n+1}_{i+1,k} - a_o \Delta q^{n+1}_{i,k} = q^*_{i,k} \tag{3.24}$$

This equation is mathematically called a *Poisson equation* (Poisson, 1813). The coefficients for uniform grid spacings are given by:

$$a_e = \Delta z/\Delta x\ ,\ a_w = \Delta z/\Delta x\ , a_t = \Delta x/\Delta z\ ,\ a_b = \Delta x/\Delta z$$

and

$$a_o = a_e + a_w + a_t + a_b$$

Figure 3.5 shows the locations of these coefficients in the Arakawa C-grid. The source term on the right-hand side of Eq. (3.24) contains the divergence of the first guess of the velocity field (u^*, w^*) and is given by:

$$q^*_{i,k} = \frac{\rho_o}{\Delta t}\left[\left(u^*_{i,k} - u^*_{i,k-1}\right)\Delta z + \left(w^*_{i,k} - w^*_{i+1,k}\right)\Delta x\right] \tag{3.25}$$

Once a solution of Eq. (3.24) is found, which implies that the new velocity field is free of divergence, the new pressure field is given by:

$$q^{n+1} = q^n + \Delta q^{n+1} \tag{3.26}$$

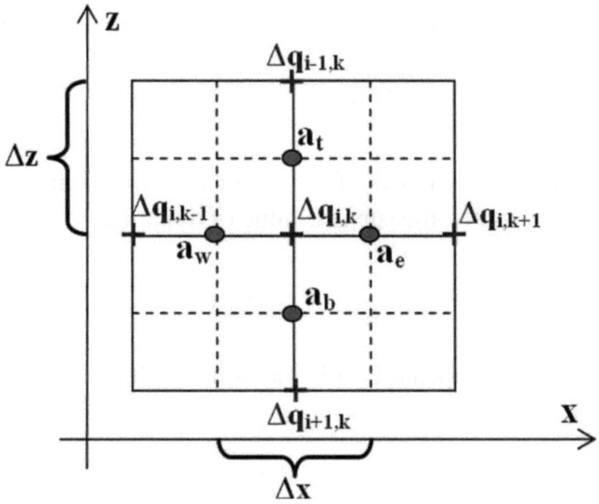

Fig. 3.5 Location used to define the coefficients a_t, a_b, a_e, and a_w. The pressure part q and the pressure correction Δq are calculated at the same grid points

3.4.4 The S.O.R. Method

The pressure equation (Eq. 3.24) can be solved by an iterative method called *Successive Over-Relaxation* (or *S.O.R* in short) that can be formulated as:

$$\Delta q_{i,k}^{r+1} = (1-\omega)\,\Delta q_{i,k}^{r} - \frac{\omega}{a_o} q_{i,k}^{*} + \\ +\frac{\omega}{a_o}\left(a_e \Delta q_{i,k+1}^{\ell} + a_w \Delta q_{i,k-1}^{\ell} + a_t \Delta q_{i-1,k}^{\ell} + a_b \Delta q_{i+1,k}^{\ell}\right) \tag{3.27}$$

where $r = 0, 1, 2, \cdots$ is the iteration index, the superscript ℓ is given by either $\ell = r+1$ or $\ell = r$ dependent on whether an update of Δq already exists, and the parameter ω determines the degree of over-relaxation. Typical values are in a range between 1.2 and 1.4.

Start values of Δq for the S.O.R. iteration can be set to zero, but the iteration is often faster if we use the values of the previous time step instead; that is,

$$\Delta q_{i,k}^{r=0} = \Delta q_{i,k}^{n}$$

The surface boundary value for dynamic pressure needs to be given at every step of the S.O.R. iteration. How this is done is described in the following. First, velocity components are updated within the S.O.R. iteration with:

$$u_{i,k}^{r+1} = u_{i,k}^{*} - \frac{\Delta t}{\rho_o \Delta x}\left(\Delta q_{i,k+1}^{r+1} - \Delta q_{i,k}^{r+1}\right) \tag{3.28}$$

$$w_{i,k}^{r+1} = w_{i,k}^{*} - \frac{\Delta t}{\rho_o \Delta z}\left(\Delta q_{i-1,k}^{r+1} - \Delta q_{i,k}^{r+1}\right) \tag{3.29}$$

The vertically integrated horizontal velocity can then be calculated from:

$$u^{r+1}_{\text{sum},k} = \sum_i (u^{r+1}_{i,k} \Delta z) \tag{3.30}$$

Slight variations of the thickness of the water column owing to sea-level elevations are ignored here. Finally, the surface value of dynamic pressure can be adjusted via:

$$\Delta q^{r+1}_{0,k} = q^{r+1}_{0,k} - q^{n}_{0,k} = -dt/\Delta x \left(u^{r+1}_{\text{sum},k} - u^{r+1}_{\text{sum},k-1}\right) \tag{3.31}$$

The S.O.R. iteration is repeated until the solution has converged to an almost steady value; that is,

$$\left|\Delta q^{r+1}_{i,k} - \Delta q^{r}_{i,k}\right| < \epsilon \tag{3.32}$$

where ϵ is a user-specified value of pressure accuracy. From Eq. (3.27), this convergence implies that:

$$a_o \Delta q^{r+1}_{i,k} \approx -q^{*}_{i,k} + \left(a_e \Delta q^{r+1}_{i,k+1} + a_w \Delta q^{r+1}_{i,k-1} + a_t \Delta q^{r+1}_{i-1,k} + a_b \Delta q^{r+1}_{i+1,k}\right) \tag{3.33}$$

which (approximately) reproduces the original Poisson equation (Eq. 3.24). Eventually, the S.O.R. method gives values of variables at the next time level $(n+1)$:

$$\begin{aligned} q^{n+1}_{i,k} &= q^{n}_{i,k} + \Delta q^{r+1}_{i,k} \\ u^{n+1}_{i,k} &= u^{r+1}_{i,k} \\ w^{n+1}_{i,k} &= w^{r+1}_{i,k} \end{aligned}$$

where the values with superscript $r+1$ are the result of the S.O.R. iteration. The flow chart in Fig. 3.6 summarises the steps that make up the S.O.R. scheme.

Surface boundary conditions are already implemented in this scheme. Boundary conditions for q at solid boundaries remain to be specified. Disappearance of vertical speed at horizontal solid surfaces implies that $\partial q/\partial z = 0$, which follows from the vertical momentum equation of Eq. (3.3). This can be implemented by setting the coefficient a_b in Eq. (3.24) to zero at such boundaries. Vertical solid boundaries are treated analogously.

The choice of the value for ϵ has implications for the accuracy of the dynamics predicted. To obtain a measure of this accuracy, we can convert the pressure accuracy ϵ into an equivalent sea-level anomaly; that is,

$$\epsilon \approx \Delta q \approx \rho_o g \Delta \eta \text{ or } \Delta \eta \approx \frac{\epsilon}{g \rho_o} \tag{3.34}$$

With a choice of $\epsilon = 0.001$ Pa, for instance, the accuracy in terms of sea-level anomalies is $\Delta\eta < 10^{-5}$ cm. This relatively high accuracy, however, can come at a cost of >500 iterations of the S.O.R. scheme for each time step of the

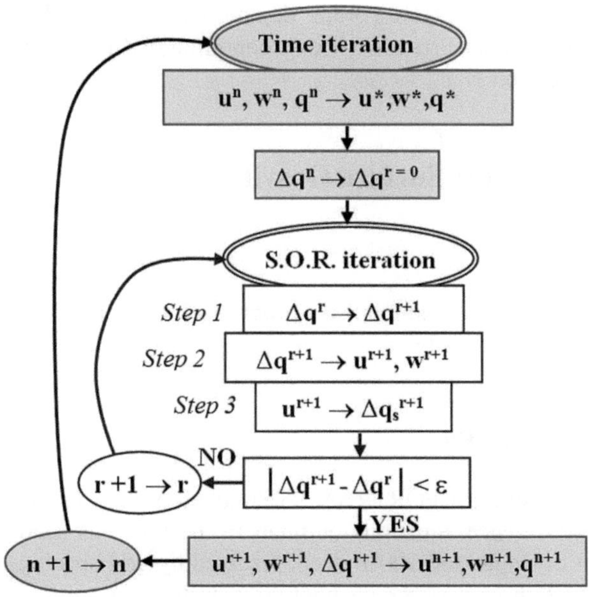

Fig. 3.6 Flow chart of the S.O.R. method

simulation. This can be rather time consuming, particularly for three-dimensional model applications.

3.4.5 Boundary Conditions for Variable Bathymetry

Variable bathymetry requires flow normal to a solid boundary to vanish. To this end, a logical array is used as to indicate whether a grid cell is "dry" or "wet". Horizontal and vertical velocity components are kept at zero values in dry grid cells during a simulation.

Owing to the staggered nature of the Arakawa-C grid (see Fig. 3.3), additional conditions are required to make sure the absence of flow across solid boundaries. This is achieved by setting horizontal velocity to zero value if the adjacent grid cell with index $(k+1)$ is dry and by setting vertical velocity to zero if the overlying grid cell of index $(i-1)$ is dry. Lateral flooding of coasts is not implemented here.

3.4.6 Stability Criterion

The stability criterion for the finite-difference versions of Eqs. (3.15, 3.16, 3.17, and 3.18) is given by:

$$\Delta t \leq \frac{\Delta x}{\sqrt{g h_{\max}}} \tag{3.35}$$

where h_{max} is the maximum total water depth encountered in the model domain. This is known as Courant-Friedrichs-Lewy condition or *CFL condition* for surface gravity waves (Courant et al., 1928).

3.5 Exercise 3: Short Surface Gravity Waves

3.5.1 Aim

The aim of this exercise is to simulate the progression of short surface gravity waves in an ocean uniform in density.

3.5.2 Task Description

The task is to construct a FORTRAN 95 simulation code based on the nonhydrostatic finite-difference equations outlined in the previous section. Consider a channel, 500 m in length and 100 m deep, resolved by grid spacings of $\Delta x = 5\,\text{m}$ and $\Delta z = 2\,\text{m}$. Both ends of the channel are closed. *Zero-gradient conditions* for dynamic pressure are applied at these boundaries. These boundary conditions imply that pressure surfaces intersect the boundary at a right angle, which is consistent with the condition of vanishing normal flow.

Forcing consists of oscillatory sea level variations of 1 m in amplitude on a period of 8 s, prescribed near the left boundary. According to the dispersion relation of short waves, given by Eq. (3.9), the expected wave length for a given period is:

$$\lambda = \frac{g}{2\pi} T^2$$

A forcing period of $T = 8\,\text{s}$ gives $\lambda \approx 100\,\text{m}$. Smart readers might complain that the resultant gravity wave is rather in the transition regime than a pure deep-water wave. Nevertheless, Eq. (3.12) suggests that the amplitude of pressure fluctuations at a depth of 100 m are only 4% compared with that at the surface. Pressure fluctuations are therefore expected to decrease rapidly with depth being the sole effect of nonhydrostatic dynamics. A time step of $\Delta t = 0.05\,\text{s}$ is chosen for adequate resolution of the forcing period. The total simulation time is 100 secs with data outputs at every second of the simulation. The author used $\omega = 1.4$ together with a pressure accuracy of $\epsilon = 0.001\,\text{Pa}$. The reader is encouraged to vary these values.

3.5.3 Results

The predicted wave pattern attains a wavelength of approximately 100 m, which is in excellent agreement with theory (Fig. 3.7). Pressure fluctuations decrease markedly with depth and vanish near the sea floor. Figure 3.7 gives the erroneous impression we are dealing with a multi-layer model. In fact, dynamic pressure is calculated

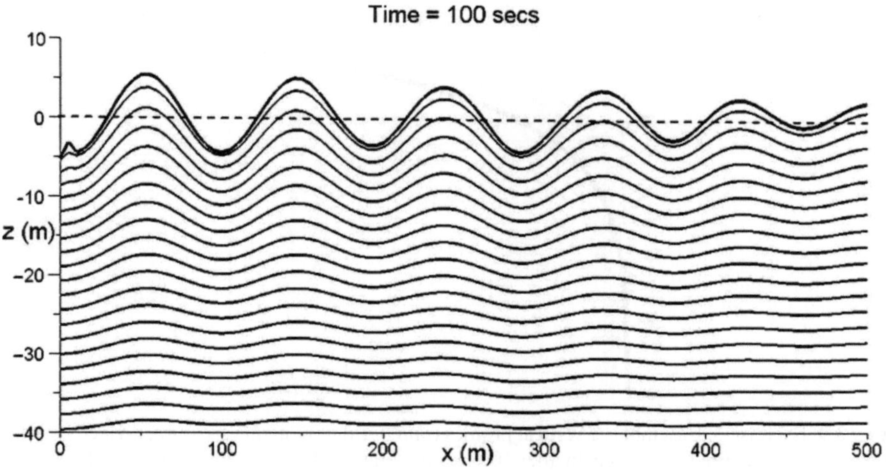

Fig. 3.7 Exercise 3. Horizontal distribution of dynamic pressure (contours), expressed by equivalent sea-level elevations ($q/(\rho_o g)$), for the upper 20 levels of the water column after 100 secs of simulation. Elevations are magnified by a factor of 5. The *top line* shows the sea surface

Fig. 3.8 Exercise 3. Dynamic pressure field, divided by $\rho_o g$, in units of metres after 80 secs of simulation. *Lines* are contours drawn at an interval of 0.1 m. Only the top 50 m of the water column is shown

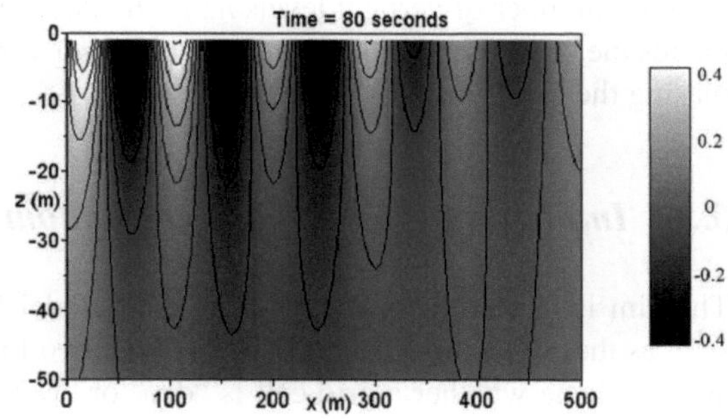

here at fixed vertical levels and the true distribution of dynamic pressure is shown in Fig. 3.8.

Theory (Eq. 3.12) suggests that the amplitude of pressure anomalies under a shot surface gravity wave decreases with depth according to:

$$q(z^*) = q_s \exp\left(-2\pi \frac{z^*}{\lambda}\right) \tag{3.36}$$

Figure 3.9 shows an excellent agreement between prediction and theory.

3.5.4 Additional Exercise for the Reader

The reader is encouraged to experiment with different forcing periods and/or total water depths and to compare the predicted wavelength of the wave with theory. Note that the dispersion relation for long waves (Eq. 3.8) gives the relationship:

$$\lambda = \sqrt{gh}T$$

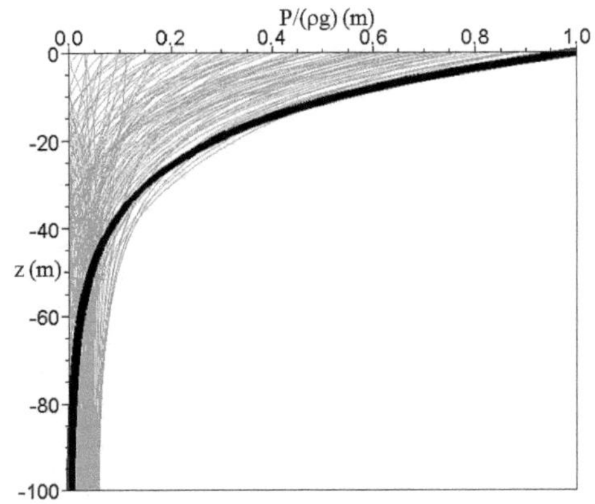

Fig. 3.9 Exercise 3. Ensemble of vertical profiles (*thin lines*) of dynamic pressure, divided by $\rho_o g$, taken from the entire model domain after 100 secs of iteration. The *thick line* shows the theoretical result of maximum pressure values according to Eq. (3.12)

Total water depth can be changed by variation of either vertical grid spacing or the total number of vertical levels nz in the declaration section. The author recommends the second option knowing that a coarser vertical resolution can act as a filter biasing the true dynamics of the process.

3.5.5 Implementation of Variable Bottom Topography

The aim is to test the vertical ocean-slice model for variable bottom topography, such as that shown in Fig. 3.10. To this end, two logical pointer arrays are declared as to indicate whether a grid cell is "dry" or "wet". Pointer values are specified at pressure grid points.

Zero-gradient conditions for dynamic pressure need to be implemented at all solid surfaces. This is done implicitly by setting the respective coefficients in the Poisson equation (Eq. 3.27) to zero and by performing the S.O.R. iteration exclusively in wet grid cells. Note that calculation of new velocity components is restricted to wet grid points excluding those defining a solid surface (see Fig. 3.3). The folder "Miscellaneous/Exercise 3 Variation" on the book's ftp site contains the modified model code.

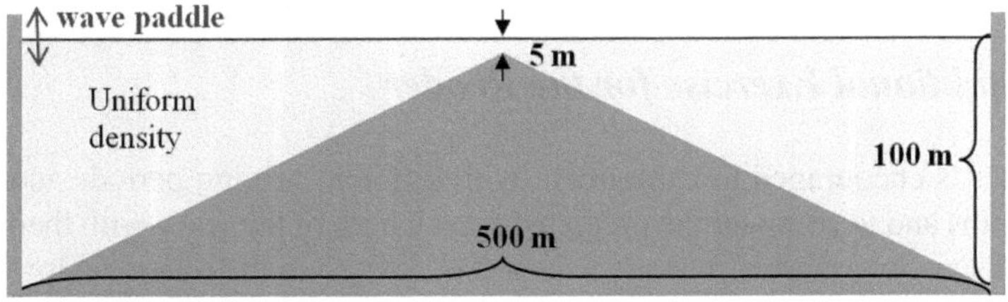

Fig. 3.10 A simple bathymetry for tests of the vertical ocean-slice model

3.5.6 Results

Figure 3.11 demonstrates that the model can also successfully cope with variable bottom topography. As the deep-water wave approaches shallower water, it turns into a shallow-water wave (because the ratio between wavelength and water depth decreases). In this region, the wave "feels" the bottom and it becomes subject to some deformation; that is, its phase speed and wavelength change in vicinity of the riff. As the wave approaches deeper water in the lee of the riff, it becomes transformed back into a deep-water wave and continues its propagation largely independent of total water depth. Note that flooding of dry land areas, such as an isolated island, is not possible with this model version.

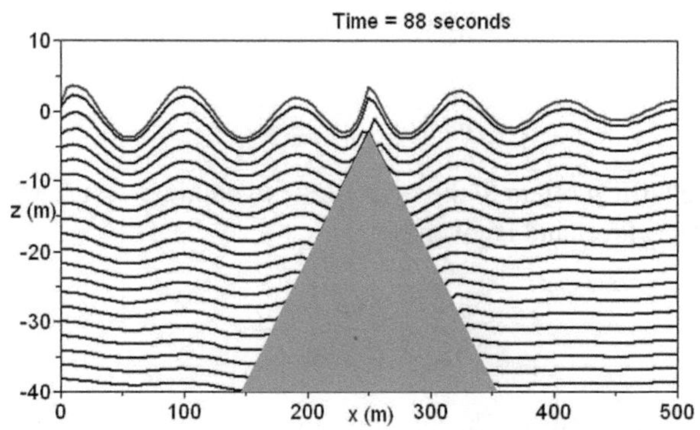

Fig. 3.11 Same as Fig. 3.7, but with variable bottom topography and after 88 secs of simulation

3.6 Inclusion of Variable Density

3.6.1 The Governing Equations

Density effects can be included in the vertical ocean-slice model by adding (a) an advection-diffusion equation for density and (b) the reduced-gravity force in the vertical momentum equation. The governing equations can then be written as:

$$\begin{aligned}
\frac{\partial u}{\partial t} + u\frac{\partial u}{\partial x} + w\frac{\partial u}{\partial z} &= -\frac{1}{\rho_o}\frac{\partial P}{\partial x} \\
\frac{\partial w}{\partial t} + u\frac{\partial w}{\partial x} + w\frac{\partial w}{\partial z} &= -\frac{1}{\rho_o}\frac{\partial P}{\partial z} - \frac{(\rho - \rho_o)}{\rho_o}g \\
\frac{\partial \rho}{\partial t} + u\frac{\partial \rho}{\partial x} + w\frac{\partial \rho}{\partial z} &= \frac{\partial}{\partial x}\left(K_h\frac{\partial \rho}{\partial x}\right) + \frac{\partial}{\partial z}\left(K_z\frac{\partial \rho}{\partial z}\right) \\
\frac{\partial u}{\partial x} + \frac{\partial w}{\partial z} &= 0 \\
\frac{\partial P_s}{\partial t} &= -\rho_o g\frac{\partial (h\langle u\rangle)}{\partial x}
\end{aligned} \tag{3.37}$$

Note that, for convenience, the author has also included the nonlinear terms in the momentum equations. By splitting pressure into parts according to Eq. (3.13), these equations take the form:

$$
\begin{aligned}
\frac{\partial u}{\partial t} + u\frac{\partial u}{\partial x} + w\frac{\partial u}{\partial z} &= -\frac{1}{\rho_o}\frac{\partial (p+q)}{\partial x} \\
\frac{\partial w}{\partial t} + u\frac{\partial w}{\partial x} + w\frac{\partial w}{\partial z} &= -\frac{1}{\rho_o}\frac{\partial q}{\partial z} \\
\frac{\partial \rho}{\partial t} + u\frac{\partial \rho}{\partial x} + w\frac{\partial \rho}{\partial z} &= \frac{\partial}{\partial x}\left(K_h\frac{\partial \rho}{\partial x}\right) + \frac{\partial}{\partial z}\left(K_z\frac{\partial \rho}{\partial z}\right) \\
\frac{\partial u}{\partial x} + \frac{\partial w}{\partial z} &= 0 \\
\frac{\partial q_s}{\partial t} &= -\rho_o g\frac{\partial (h\,\langle u\rangle)}{\partial x}
\end{aligned}
\tag{3.38}
$$

The pressure term p refers to the hydrostatic dynamic pressure with reference to an undisturbed sea level and is given by:

$$
\frac{\partial p}{\partial z} = -\frac{(\rho - \rho_o)}{\rho_o} g \tag{3.39}
$$

with $p = 0$ at the sea surface. This ignores effects due to atmospheric pressure variations, which usually can be neglected. Notice that the reduced-gravity force (often symbolised in short as g') has disappeared from the vertical momentum equation, but, in fact, this force has been shifted into the hydrostatic pressure part and density effects appear now in the horizontal momentum equations.

3.6.2 Discretisation of the Advection Terms

The advection equation for a variable B subject to flow with components u and w in the vertical ocean slice is given by:

$$
\frac{\partial B}{\partial t} = -u\frac{\partial B}{\partial x} - w\frac{\partial B}{\partial z} \tag{3.40}
$$

Using the product rule of differentiation, this equation can be reformulated as:

$$
\frac{\partial B}{\partial t} = -\frac{\partial (uB)}{\partial x} - \frac{\partial (wB)}{\partial z} + B\left(\frac{\partial u}{\partial x} + \frac{\partial w}{\partial z}\right) \tag{3.41}
$$

where the last term vanishes with insertion of the continuity equation (Eq. 3.17).

The temporal change of B can be discretised using a simple time-forward iteration. The treatment of the remaining term is described in the following for the x-direction. Analog recipes apply for the z-direction. The general approach is to consider a control volume (Fig. 3.12) and to express fluxes of B through the faces of this control volume as:

$$-\Delta t\frac{\partial(uB)}{\partial x} = C_w B_w - C_e B_e \tag{3.42}$$

where the indices "w" and "e" refer to east and west faces of the control volume, and the C parameters are so-called *Courant numbers*. For variables located at pressure grid points (see Fig. 3.3), for instance, the Courant numbers are given by:

$$C_w = u_{k-1}^n \Delta t/\Delta x \text{ and } C_e = u_k^n \Delta t/\Delta x$$

In a next step, u is split into positive and negative components:

$$u^+ = 0.5(u + |u|) \quad \text{and} \quad u^- = 0.5(u - |u|)$$

so that Eq. (3.42) can be rewritten as:

$$-\Delta t\frac{\partial(uB)}{\partial x} = C_w^+ B_w^+ + C_w^- B_w^- - C_e^+ B_e^+ - C_e^- B_e^-$$

On the basis of *Total Variation Diminishing* (TVD) advection schemes (see Fringer et al., 2005), the face values of B are computed with the upstream values

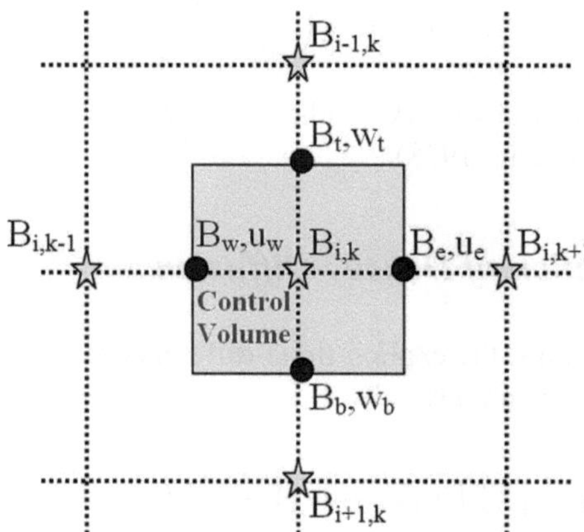

Fig. 3.12 The control volume

plus the addition of a higher-order term according to:

$$\begin{aligned}
B_e^+ &= B_k^n + 0.5\Psi\left(r_k^+\right)\left(1 - C_e^+\right)\left(B_{k+1}^n - B_k^n\right)\\
B_e^- &= B_{k+1}^n - 0.5\Psi\left(r_k^-\right)\left(1 + C_e^-\right)\left(B_{k+1}^n - B_k^n\right)\\
B_w^+ &= B_{k-1}^n + 0.5\Psi\left(r_{k-1}^+\right)\left(1 - C_w^+\right)\left(B_k^n - B_{k-1}^n\right)\\
B_w^- &= B_k^n - 0.5\Psi\left(r_{k-1}^-\right)\left(1 + C_w^-\right)\left(B_k^n - B_{k-1}^n\right)
\end{aligned}$$

where the r parameters are defined by:

$$r_k^+ = \frac{B_k^n - B_{k-1}^n}{B_{k+1}^n - B_k^n} \quad \text{and} \quad r_k^- = \frac{B_{k+2}^n - B_{k+1}^n}{B_{k+1}^n - B_k^n}$$

Here, we use the so-called Superbee scheme in which the limiting function Ψ is defined by:

$$\Psi(r) = \max\left\{0, \min(2r, 1), \min(r, 2)\right\}$$

Kämpf (2009) shows performance tests of other limiting functions. Vertical advection is discretised via fluxes of B through vertical faces of the control volume in a similar fashion (not shown here). The last term in (3.41), although it should be zero in theory, is included in this scheme to avoid accumulation of small but persistent round-off errors that could lead to artificial and unwanted internal sources or sinks of B.

3.6.3 Stability Criterion for the Advection Equation

The stability criterion for the above explicit form of the advection equation is given by:

$$\Delta t \leq \min\left(\frac{\Delta x}{u}, \frac{\Delta z}{w}\right) \tag{3.43}$$

known as Courant-Friedrichs-Lewy condition or *CFL* condition for explicit advection schemes (Courant et al., 1928).

3.6.4 Implementation of Density Diffusion

For uniform grid spacings, the explicit finite-difference versions of the density diffusion terms in Eq. (3.38) are given by:

$$\begin{aligned}
\frac{\partial}{\partial x}\left(K_h\frac{\partial\rho}{\partial x}\right) &= \left[K_h^e\left(\rho_{i,k+1} - \rho_{i,k}\right) - K_h^w\left(\rho_{i,k} - \rho_{i,k-1}\right)\right]/\left(\Delta x\right)^2\\
\frac{\partial}{\partial z}\left(K_z\frac{\partial\rho}{\partial z}\right) &= \left[K_z^+\left(\rho_{i-1,k} - \rho_{i,k}\right) - K_z^-\left(\rho_{i,k} - \rho_{i+1,k}\right)\right]/\left(\Delta z\right)^2
\end{aligned}$$

where the interpolated values of eddy diffusivity are given by:

$$K_h^e = 0.5\left(K_{h,i,k} + K_{h,i,k+1}\right) \text{ and } K_h^w = 0.5\left(K_{h,i,k} + K_{h,i,k-1}\right)$$
$$K_z^+ = 0.5\left(K_{z,i,k} + K_{z,i-1,k}\right) \text{ and } K_z^- = 0.5\left(K_{z,i,k} + K_{z,i+1,k}\right)$$

Figure 3.13 shows the locations at which these values are determined. Vanishing diffusive fluxes across solid boundaries require the use of zero-gradient conditions for density.

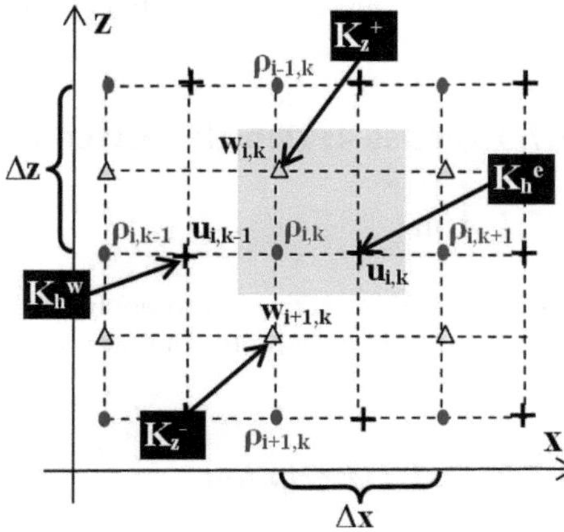

Fig. 3.13 Arakawa-C grid. Eddy viscosity is calculated at grid points where also pressure, density and other scalars are calculated. The *gray square* indicates a selected grid cell

3.6.5 Required Modifications of the Code

Several additions to the previous FORTRAN 95 simulation code are required to include effects associated with variable density. These are:

- Addition of an advection-diffusion equation for density including boundary conditions,
- Calculation of the baroclinic pressure term p from the density distribution.

As the reader might have noticed, dynamic pressure is now composed of three parts:

$$P \Rightarrow p^n + q^n + \Delta q^{n+1} \tag{3.44}$$

where p is diagnosed from:

$$p_{i,k}^n = p_{i-1,k}^n + (\langle\rho^n\rangle - \rho_o)g\Delta z \text{ for } i = 1, 2, 3, \cdots, nz$$

with $p_{0,k}^n = 0$. Density ρ is hereby interpolated to represent values centred between vertically neighboring pressure grid points. This interpolated value is given by:

$$\langle\rho^n\rangle = 0.5(\rho_{i-1,k}^n + \rho_{i,k}^n)$$

With inclusion of the nonlinear terms (required for subsequent exercises), the first-guess value of velocity is now calculated from:

$$u^*_{i,k} = u^n_{i,k} - \Delta t \mathrm{Adv}(u) - \frac{\Delta t}{\rho_o \Delta x}(p^n_{i,k+1} - p^n_{i,k} + q^n_{i,k+1} - q^n_{i,k}) \quad (3.45)$$

$$w^*_{i,k} = w^n_{i,k} - \Delta t \mathrm{Adv}(w) - \frac{\Delta t}{\rho_o \Delta z}(p^n_{i-1,k} - p^n_{i,k} + q^n_{i-1,k} - q^n_{i,k}) \quad (3.46)$$

where Adv(u) and Adv(w) represent the nonlinear terms.

3.7 Exercise 4: Density-Driven Flows

3.7.1 Aim

The aim of this exercise is to apply the vertical ocean-slice model in a study of bottom-arrested density-driven flows over variable bottom topography.

3.7.2 Task Description

Consider a closed channel, 500 m long and 100 m deep, resolved by grid spacings of $\Delta x = 5$ m and $\Delta z = 2$ m. This configuration is the same as in Exercise 3. The model is forced via prescription of a layer of dense water that initially leans against the left boundary, as shown in Fig. 3.14. This layer is initially 100 m thick and 50 m wide. Its density is 1 kg/m^3 greater compared with ambient water having a density of $\rho_o = 1{,}028$ kg/m^3. Owing to initially unbalanced lateral pressure gradients, this layer will spread along the sea floor with the aim to achieve a final state at rest void of any horizontal density gradients.

Horizontal and vertical density diffusivities are set to small uniform values of $K_h = K_z = 1 \times 10^{-4}$ m^2/s. The total simulation time is 50 min with data outputs every 30 secs. The author used a time step of $\Delta t = 0.1\,s$, which satisfies the CFL

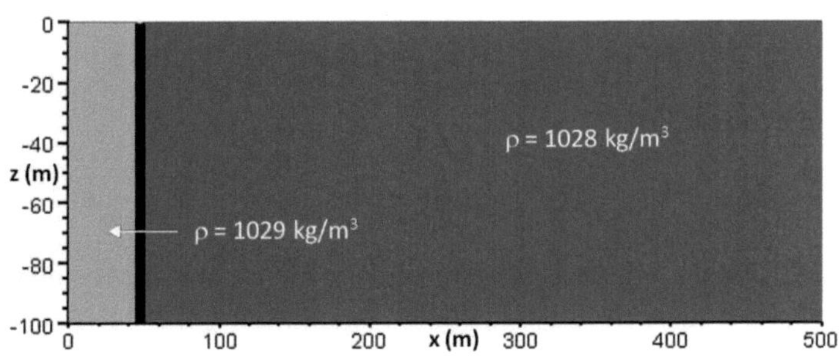

Fig. 3.14 Initial density field for Exercise 4

criteria for both external gravity waves (Eq. 3.35) and advection (Eq. 3.43). Parameters of the S.O.R. scheme are kept at the same values as in the previous exercise.

3.7.3 Theory

Baroclinic pressure gradients associated with horizontal density differences will produce a bottom-arrested density-driven flow. The resultant speed of the flow can be estimated from energy conservation principles. In this process, potential energy available from the initial density field is converted into kinetic energy of the flow. This energy conversion can be quantified by the *Bernoulli equation*. Provided that the plume density does not change, the Bernoulli equation reads:

$$0.5u^2 + g'h = g'h_o \tag{3.47}$$

where g' is reduced gravity, and h_o and h, respectively, are initial and final plume thicknesses. The Bernoulli equation is named after Daniel Bernoulli (Bernoulli, 1738) and his father Johann. The first true Bernoulli equation, however, was derived by Euler (1755). The latter equation can be rewritten as:

$$u = \sqrt{2g'(h_o - h)} \tag{3.48}$$

3.7.4 Results

As anticipated, the initial density anomaly produces a density-driven flow spreading toward the right-hand side of the model domain (Fig. 3.15). This flow forms an isolated plume head. Counter-clockwise vortices forming in the lee of this head induce vigorous mixing.

With $g' = 0.0095\,\mathrm{m\,s^{-2}}$, $h_o = 100\,\mathrm{m}$, and $h \approx 40\,\mathrm{m}$, the Bernoulli equation (Eq. 3.48) suggests a plume speed of $u = 1.1\,\mathrm{m/s}$. In good agreement with theory, maximum speeds of the simulated plume vary by $\pm 0.2\,\mathrm{m/s}$ around this value.

It should be highlighted that the model appears capable of simulating turbulence initiated by vertical shear of the horizontal flow and the breaking of internal waves. Exercise 6 will explore this feature in greater detail.

3.7.5 Can Reduced-Gravity Plumes Jump?

The author decided to repeat the latter experiment with inclusion of variable bottom topography with a ramp and a vertical bar (Fig. 3.16). Will the reduced-gravity plume make it over these obstacles?

Results show that the density-driven current is energetic enough to pass the ramp. It shoots upward as it meets the vertical bar where it forms a counter-rotating vortex

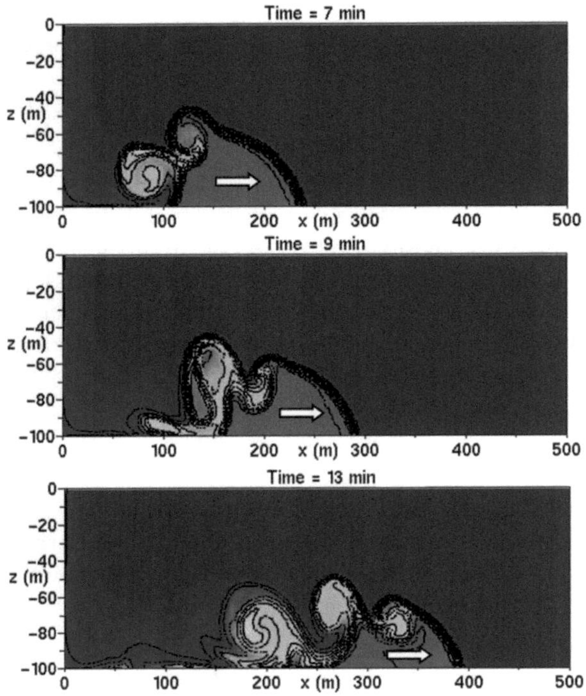

Fig. 3.15 Exercise 4: Snapshots of the density distribution (*shading* and *contours*) at selected times of the simulation

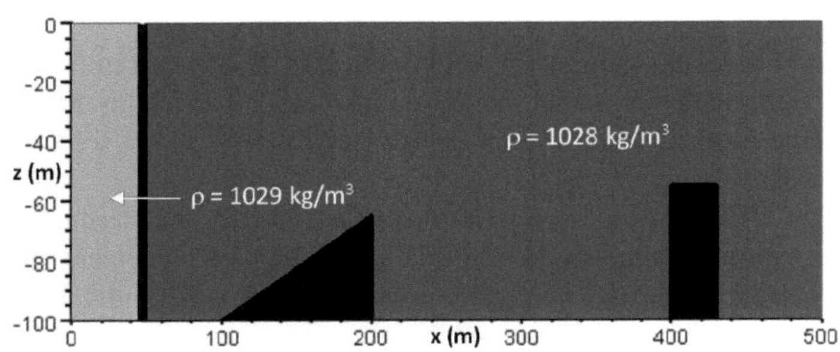

Fig. 3.16 Variation of Exercise 4. Initial density field and bathymetry

pair, sometimes called a "mushroom" (Fig. 3.17). Notice that a fraction of dense water flows back downward on the ramp.

Surely, neither the initial density distribution nor our bottom topography can be found in nature, but it is fun to create such fictional scenarios, isn't it? When I showed the GIF animation of the results in the classroom, students asked whether I believed these results were realistic. Well, we really can't tell without any means of comparison with the real situation, can we? Nevertheless, the author is confident that laboratory-based experiments would show similar results.

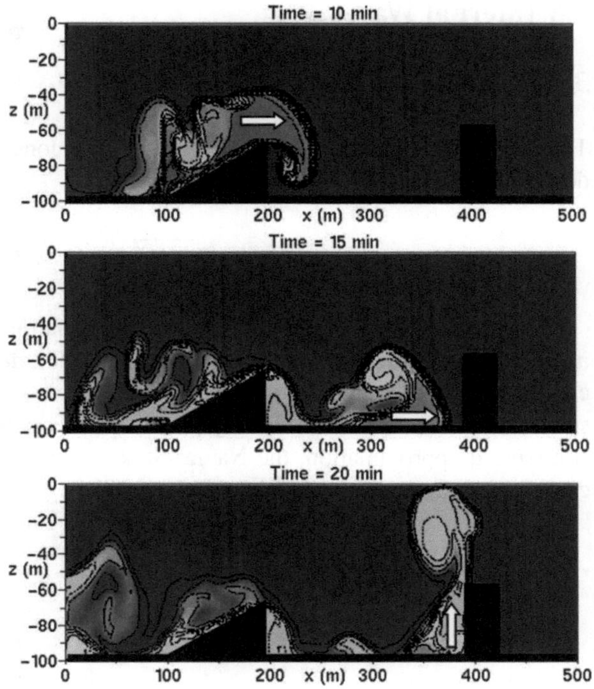

Fig. 3.17 Variation of Exercise 4: Snapshots of density distribution at selected times of the simulation

3.7.6 Additional Exercise for the Reader

Repeat this exercise, but make the forcing layer lighter compared with the ambient fluid. Can you guess what happens? Are you curious? The reader is also encouraged to run the model with use of the rigid-lid approximation being described in the following.

3.7.7 The Rigid-Lid Approximation

Inclusion of a free sea surface requires the choice of fairly small time steps to satisfy the CFL criterion associated with propagation of surface gravity waves (Eq. 3.35). Longer time steps are possible under the assumption that vertical velocity vanishes at the sea surface; that is, $w_s = 0$. This assumption can be called *rigid-lid approximation* for nonhydrostatic models. It eliminates surface gravity waves and is easily implemented in the code by setting the vertical velocity at the surface to zero via the logical wet/dry pointer array.

With use of the rigid-lid approximation, the completion time of the previous exercise can be reduced fivefold via choice of a larger time step of $\Delta t = 1$ s without major changes in the results. Nevertheless, the rigid-lid approximation might not work in other model applications and, therefore, it should not be the first choice.

3.8 Internal Waves

3.8.1 Theory

The density field in a fluid can be expressed by a constant part, a time-independent depth-variable part, and fluctuations according to:

$$\rho = \rho_o + \overline{\rho(z)} + \rho'(x, z, t) \tag{3.49}$$

Analytical solutions of the Navier-Stokes equations uncovering internal gravity waves can be derived under the assumption that density fluctuations are small compared with the depth-variable part; that is, $|\rho'| << |\overline{\rho}|$ with both parts being small compared with mean density ρ_o. Under this condition, which is a modified Boussinesq approximation, the Navier-Stokes equations for a vertical ocean slice can be approximated by (Cushman-Roisin, 1994):

$$\frac{\partial u}{\partial t} = -\frac{1}{\rho_o}\frac{\partial P}{\partial x} \tag{3.50}$$

$$\frac{\partial w}{\partial t} = -\frac{1}{\rho_o}\frac{\partial P}{\partial z} - \frac{\rho'}{\rho_o}g \tag{3.51}$$

$$\frac{\partial u}{\partial x} + \frac{\partial w}{\partial z} = 0 \tag{3.52}$$

$$\frac{\partial \rho}{\partial t} + w\frac{\partial \overline{\rho}}{\partial z} = 0 \tag{3.53}$$

where higher-order terms associated with diffusion and advection are ignored. In the following we also assume that $\overline{\rho(z)}$ varies linearly with depth corresponding to a constant stability frequency: $N^2 = -g/\rho_o \partial\overline{\rho}/\partial z$.

In a first consideration, the existence of a sea surface is ignored, assuming an ocean of infinite vertical extent. For sinusoidal linear waves that are allowed to travel into any direction of the vertical ocean slice, it can be shown that the frequency of the wave σ obeys the dispersion relation (e.g. Cushman-Roisin, 1994):

$$\sigma = \frac{2\pi}{T} = \pm N \cos\theta$$

where T is wave period and θ is the angle between the direction of wave propagation and the horizontal plane (Fig. 3.18). Hence, the minimum period an internal wave in an ocean of continuous density stratification can attain is:

$$T_{\min} = \frac{2\pi}{N} \tag{3.54}$$

which implies purely horizontal propagation. This corresponds to the period of oscillations experienced by a buoyant object in a stratified fluid (see Sect. 3.10 of Kämpf (2009)). Consequently, such internal waves propagate at a phase speed of:

$$c = \frac{\lambda}{T} = \pm\frac{\lambda N \cos(\theta)}{2\pi}$$

where λ is the wavelength. The fact that two signs are allowed indicates that the wave can travel into one of two directions (see Fig. 3.18). If the frequency of an internal wave is imposed via external forcing, regardless of wavelength, all waves propagate at a certain fixed angle from the horizontal. The longer the period the steeper the direction.

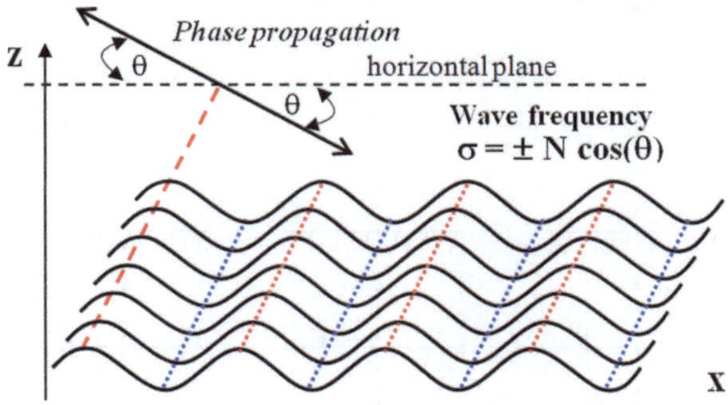

Fig. 3.18 Vertical structure of an internal wave. Adapted from Cushman-Roisin (1994)

3.8.2 Normal Wave Modes

In the ocean interior, internal wave motion can induce large vertical excursions of density interfaces of several tens of metres. Vertical boundaries (sea surface and sea floor) do not permit such large-amplitude vertical oscillations. Consequently, vertical velocity inherent with internal waves has to vanish at the sea floor and it has to become very small at the sea surface. Since the resultant waves can only propagate horizontally, the vertical boundaries operate as a waveguide.

Only a discrete set of wave solutions, so-called *normal modes*, satisfies the conditions of vanishing vertical velocity at vertical boundaries. It can be shown that, for a constant stability frequency N, possible wave frequencies are (Pond and Pickard, 1983):

$$\sigma = \pm\frac{N}{\sqrt{1 + (0.5n\lambda/h)^2}}$$

where $n = 1, 2, 3, \cdots$ is the mode number, λ is horizontal wavelength, and h is total water depth.

Figure 3.19 displays the vertical structure of horizontal and vertical velocity components of the first three internal wave modes. The mode number n gives the number of maxima of the vertical velocity profile and the number of nodes (zero crossings) of the horizontal speed profile. Phase speeds associated with individual wave modes are given by:

$$c = \frac{\lambda}{T} = \pm \frac{N}{\sqrt{(2\pi/\lambda)^2 + (n\pi/h)^2}} \tag{3.55}$$

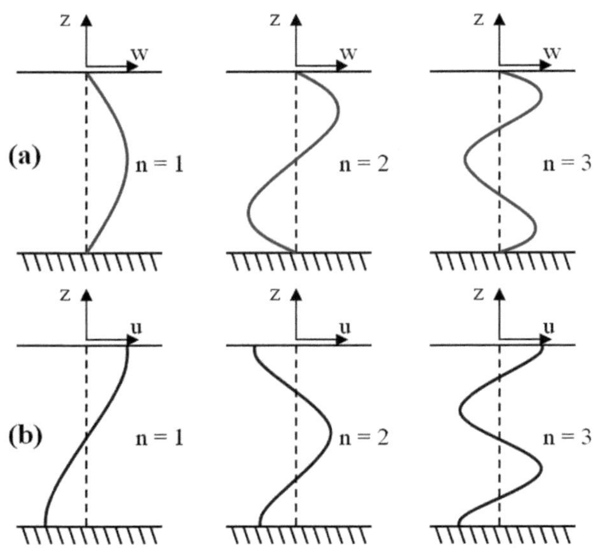

Fig. 3.19 Vertical structure of (**a**) *vertical* and (**b**) *horizontal* velocity components for the first three normal modes of internal waves with N = constant. Adapted from Pond and Pickard (1983)

3.9 Exercise 5: Internal Waves

3.9.1 Aim

The aim of this exercise is to simulate the dynamics of internal waves in a finite-depth ocean of continuous density stratification on the basis of Eqs. (3.38).

3.9.2 Task Description

Consider a model domain of 500 m in length and 100 m in depth, resolved by grid spacings of $\Delta x = 5$ m and $\Delta z = 2$ m. Both lateral boundaries are closed. To make the dynamics slightly more complex, the author decided to add a shallower region

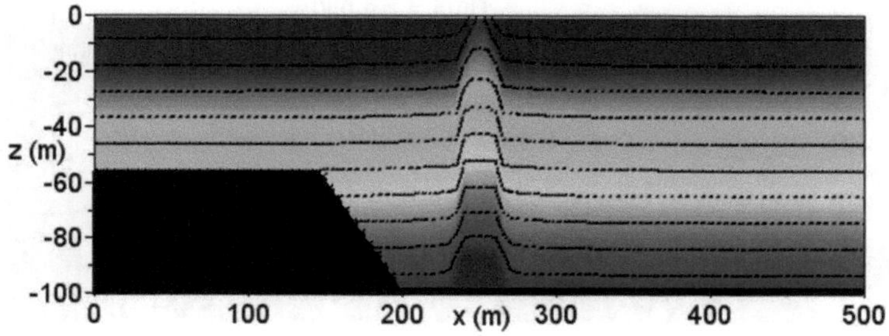

Fig. 3.20 Initial density distribution (*shading* and *contours*) and bathymetry for Exercise 5

near the left boundary (Fig. 3.20). The ambient ocean is initially at rest and linearly stratified characterised by a stability frequency of $N = 0.1\ \text{s}^{-1}$. Surface density is set to $\rho_s = 1{,}028\,\text{kg/m}^3$. The minimum period of internal waves according to Eq. (3.54) is about 63 s. Such a strong density stratification does rarely exist in the ocean. The sole purpose here is to minimise the total simulation time whilst capturing about 10 wave periods.

An initial density disturbance is added in the centre of the model domain over a width of 25 m (five adjacent grid cells). Density is increased by $20\,\text{kg/m}^3$ in the water column in this region, but the maximum density is limited by the bottom density found in ambient water. The purpose of this treatment is to prevent creation of a density-driven bottom-arrested flow which already has been studied in the previous exercise.

Horizontal and vertical density diffusivities are set to small uniform values of $K_h = K_z = 1 \times 10^{-4}\ \text{m}^2/\text{s}$. The rigid-lid version of the nonhydrostatic vertical ocean-slice model is applied with a time step of $\Delta t = 1\,\text{s}$ and a pressure accuracy of $\epsilon = 0.01\,\text{Pa}$. The simulation time is 10 mins with data outputs every 10 secs.

3.9.3 Results

Owing to reflection at closed boundaries, wave disturbances in closed domains trigger the formation of *standing waves*. Standing waves are waves of zero horizontal propagation that, at certain locations called nodes, exhibit no vertical displacements. Exclusively vertical displacements are found between nodes and near lateral boundaries. The initial density disturbance creates an internal wave pattern that becomes reflected at the closed lateral boundaries (Fig. 3.21). At times, internal waves break and induce vertical mixing. Reflection at closed boundaries gives rise to a complex wave pattern. Vertical profiles of vertical velocity (Fig. 3.22) reveal wave shapes of one or two maxima in the water column corresponding to mode numbers $n = 1$ and $n = 2$ (see Fig. 3.19), demonstrating that the vertical boundaries of fluids operate as a waveguide. Notice that the vertical speed of wave motions exceeds 40 cm/s.

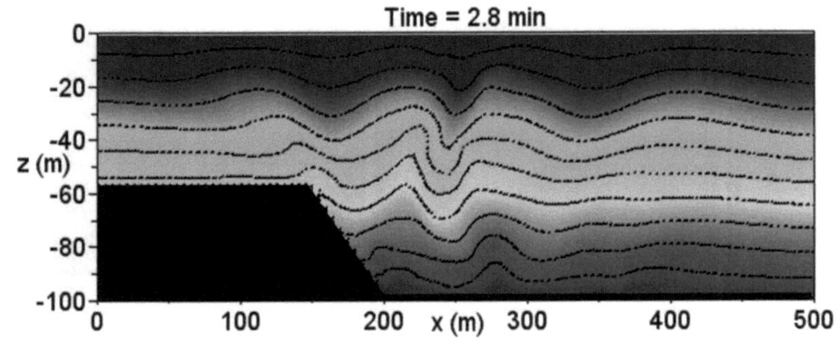

Fig. 3.21 Exercise 5. Density distribution after of 2.8 min of simulation

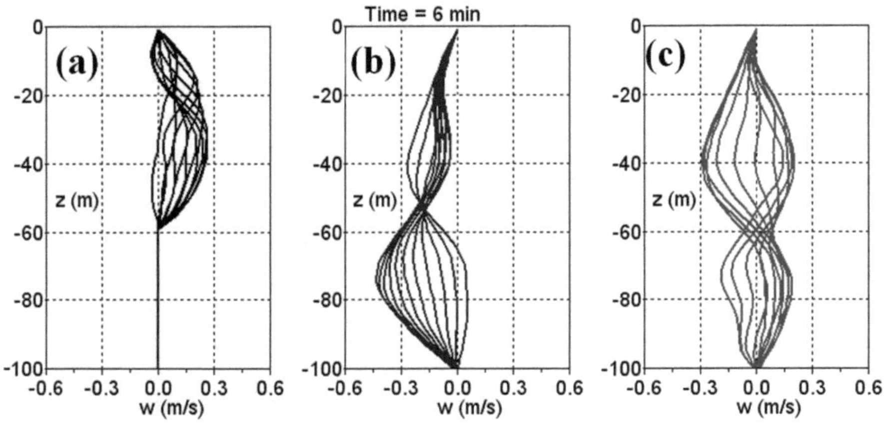

Fig. 3.22 Exercise 5. Ensemble of vertical profiles of vertical velocity after 6 min of iteration for the zones (**a**) $x = 50$–100 m, (**b**) $x = 350$–400 m, and (**c**) $x = 400$–450 m

3.9.4 Additional Exercise for the Reader

Place the initial density anomaly in the shallower region and explore the resultant wave patterns.

3.10 Mechanical Turbulence

3.10.1 Kelvin-Helmholtz Instability

Shear flows are flows of a speed that varies in a direction perpendicular to the current. This can be either in the horizontal plane where the resultant flows are referred to as horizontal shear flows and/or in the vertical plane characterising vertical shear flows. The latter are considered in the following. Vertical shear flows tend to create turbulence inducing mixing in the water column. In a stratified fluid, work has to

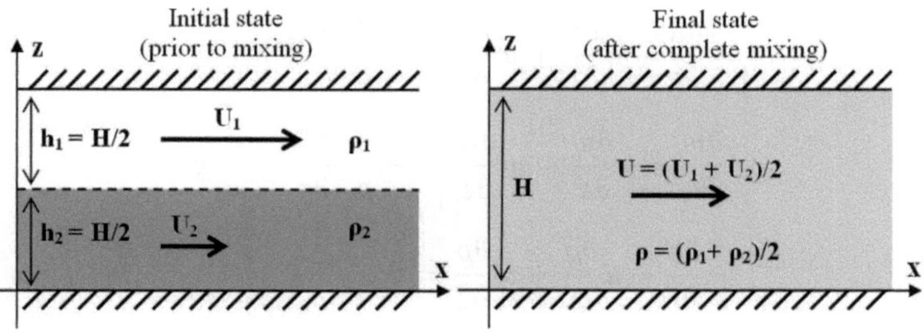

Fig. 3.23 Mixing of a two-layer stratified fluid with flow shear. Adapted from Cushman-Roisin (1994)

be performed to raise heavier fluid parcels and to lower lighter fluid parcels against the reduced-gravity force. Hence, density stratification generally operates to lower turbulence levels.

Let us consider a simplified system pictured in Fig. 3.23. Initially, there are two superimposed layers of different densities in a stable stratification (lighter water on top of heavier water). Each layer is in motion at a certain uniform speed, but there is a strong velocity shear across the density interface. With equal layer thicknesses, energy conservation suggests that complete mixing occurs for (Cushman-Roisin, 1994):

$$\frac{(\rho_2 - \rho_1)\, gH}{\rho_o\, (U_2 - U_1)^2} < 1$$

where $\rho_o \approx \rho_1 \approx \rho_2$ is a mean density, using the Boussinesq approximation. Localised mixing in vicinity of the density interface, however, is possible, if the wavelength of a perturbation λ is such that (Kundu, 1990):

$$\frac{\pi\, (\rho_2 - \rho_1)\, g}{\rho_o\, (U_2 - U_1)^2}\lambda < 1 \quad \text{or} \quad \lambda < \frac{\rho_o}{|\Delta\rho|}\frac{(\Delta u)^2}{\pi g} \tag{3.56}$$

We can therefore anticipate that there are always sufficiently short waves to create instability. Therefore, a two-layer shear flow is always unstable. This is known as the *Kelvin-Helmholtz instability*, first described by Thomson (Lord Kelvin) (1871) and Helmholtz (1868). Incidentally, this mechanism explains the generation of water waves by surface winds.

3.10.2 Instability of a Stratified Shear Flow

Analytical solutions of the shear-flow problem can be derived under the assumptions that both the horizontal flow and density vary gradually with depth. The dynamics of this problem in a vertical ocean slice slice can be approximated by the set of equations:

$$\frac{\partial u}{\partial t} + u\frac{\partial u}{\partial x} + w\frac{\partial u}{\partial z} = -\frac{1}{\rho_o}\frac{\partial P}{\partial x} \tag{3.57}$$

$$\frac{\partial w}{\partial t} + u\frac{\partial w}{\partial x} + w\frac{\partial w}{\partial z} = -\frac{1}{\rho_o}\frac{\partial P}{\partial z} - \frac{\rho'}{\rho_o}g \tag{3.58}$$

$$\frac{\partial \rho}{\partial t} + u\frac{\partial \rho}{\partial x} + w\frac{\partial \rho}{\partial z} = 0 \tag{3.59}$$

$$\frac{\partial u}{\partial x} + \frac{\partial w}{\partial z} = 0 \tag{3.60}$$

where density diffusion and frictional effects are ignored, to zero-order approximation. Also, the existence of the sea surface is ignored, based on the assumption of a fluid of infinite vertical extent. For a constant stability frequency N and an initial horizontal flow of a speed that varies linearly with depth ($\partial u/\partial z$ = constant), it can then be shown that instability occurs locally, if the so-called Richardson number falls below a threshold value of 1/4 (Richardson, 1920). The Richardson number is hereby defined by:

$$Ri = \frac{N^2}{(\partial u/\partial z)^2} \tag{3.61}$$

Accordingly, instability in a two-layer shear fluid, such as that depicted in Fig. 3.23, can be expected to produce a final transition layer of thickness Δh with $Ri \approx 0.25$ throughout this layer. Using Eq. (3.61), the thickness of the resultant transition layer can be estimated from the equation:

$$\frac{-g/\rho_o \Delta\rho/\Delta h}{(\Delta u/\Delta h)^2} = 0.25 \quad \text{or} \quad \Delta h = 0.25\frac{\rho_o}{|\Delta\rho|}\frac{(\Delta u)^2}{g} \tag{3.62}$$

where $\Delta\rho$ and Δu, respectively, are differences in density and flow speed across this layer. We will use the latter equation for verification of simulation results. Note that the vertical scale from Eq. (3.62) is of the same order of magnitude as the horizontal wavelength described by Eq. (3.56). Hence, horizontal and vertical length scales of turbulent vortices inherent with the Kelvin-Helmholtz instability mechanism are closely related to each other, which implies a nonhydrostatic nature of the dynamics at play.

3.11 Exercise 6: Kelvin-Helmholtz Instability

3.11.1 Aim

The aim of this exercise is to simulate mixing at the density interface of a two-layer vertical shear fluid.

3.11.2 Task Description

Consider a vertical ocean slice of 500 m in length and 100 m in depth with cyclic lateral boundaries (see below) and grid spacings of $\Delta x = 5\,\text{m}$ and $\Delta z = 2\,\text{m}$. The density interface of two superimposed layers of different densities is located at a depth of 50 m. The density of the top layer is set to $\rho_1 = 1{,}027.0\,\text{kg/m}^3$. The bottom layer has a density of $\rho_2 = 1{,}029.0\,\text{kg/m}^3$, yielding a density difference of $\Delta\rho = 2\,\text{kg/m}^3$. Density diffusion is included in the model. Horizontal and vertical density diffusivities are set to small uniform values of $K_h = K_z = 1 \times 10^{-4}\,\text{m}^2/\text{s}$.

The initial lateral flow in the top layer has an average speed of 1 m/s, whereas the lower layer is at rest. Randomly generated values of ± 0.01 m/s are added to the upper-layer flow. This technique promotes the initial growth of disturbances at random locations. The random-number generator was taken from Press et al. (1989). With a difference in layer speeds of $\Delta u = 1\,\text{m/s}$, Eq. (3.62) suggests that the transition mixing zone attains a thickness of around 13 m. Equation (3.56) predicts that mixing is initiated by disturbances of <18 m in wavelength. These length scales are barely resolved with the grid spacings chosen. The reader is encouraged to use a finer spatial resolution for more accurate results.

This task employs the rigid-lid version of the vertical ocean-slice model. The total simulation time is 100 min with outputs every minute. The time step is set to $\Delta t = 1\,\text{s}$. The pressure accuracy for the S.O.R. iteration is set to $\epsilon = 1 \times 10^{-3}$ Pa. Will the model adequately simulate vertical mixing inherent with the Kelvin-Helmholtz instability mechanism?

3.11.3 Cyclic Boundary Conditions

Cyclic boundary conditions mean that opposite lateral boundaries are connected with each other, such that flow escaping through one boundary enters through the opposite one. With our numerical grid, cyclic boundary conditions are implemented with:

$$\psi_{k=0} = \psi_{k=nx}$$
$$\psi_{k=nx+1} = \psi_{k=1}$$

where ψ stands for any variable. The use of cyclic boundary conditions is justified for dynamical disturbances of a wavelength much shorter than the length of the model domain and occupying the entire model domain. Cyclic boundary conditions correspond then to a model domain of infinite length. A rule of thumb is that the model domain should extend at least ten wavelengths of disturbances. Otherwise, some stretching bias will occur.

3.11.4 Results

The flow becomes dynamically unstable owing to the Kelvin-Helmholtz instability mechanism after 45 min of simulation and creates vigorous internal wave breaking at the density interface (Fig. 3.24). Disturbances attain vertical speeds of $> 30\,\text{cm/s}$. Equation (3.60) suggests that disturbances are confined to a layer of about 13 m in thickness, such that $Re \approx 0.25$ establishes in this layer. The simulation produces a transition zone of around 25 m in thickness, characterised by a Richardson number of $Ri \approx 0.5$ (Figs. 3.25 and 3.26). This discrepancy by a factor of two is presumably caused by inertia effects followed by continued mixing after onset of dynamical instabilities. Another source of bias could be the relatively coarse grid spacing, which the reader may verify via the choice of finer grid spacings.

Vortices involved in the instability process do neither fully mix density nor momentum. The final result is rather a transition zone over which both density and the horizontal flow vary approximately linearly. Hence, it is a misconception to assume that the Kelvin-Helmholtz mechanism fully mixes portions of the fluid column. In model applications that cannot resolve the Kelvin-Helmholtz instability mechanism, this process is often parameterised by means of vertical turbulent diffusion in which the coefficient is a function of the Richardson number.

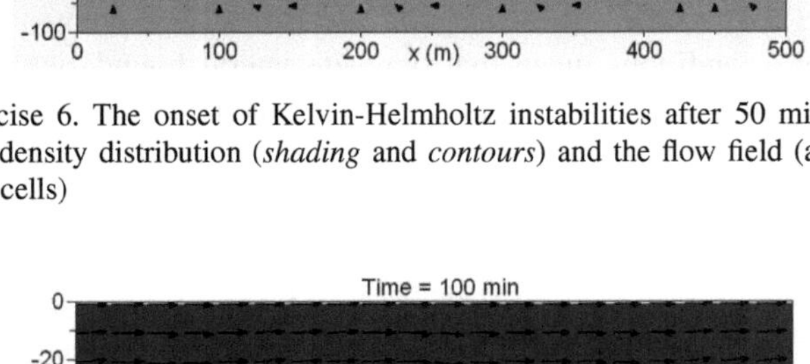

Fig. 3.24 Exercise 6. The onset of Kelvin-Helmholtz instabilities after 50 min of simulation. Shown are the density distribution (*shading* and *contours*) and the flow field (arrows, averaged over 5 × 5 grid cells)

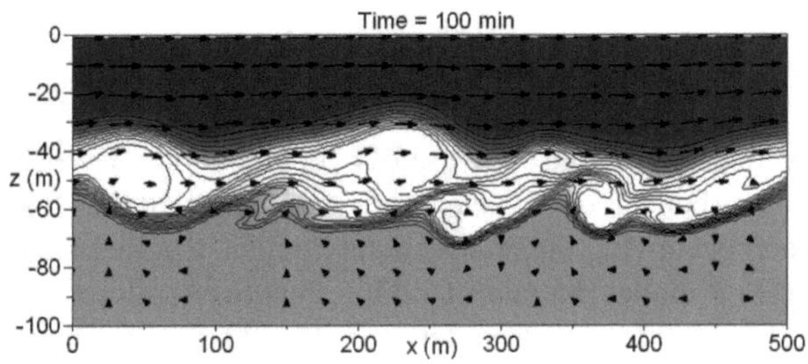

Fig. 3.25 Exercise 6. Same as Fig. 3.24, but after 100 min of simulation

3.11.5 Additional Exercise for the Reader

Repeat this exercise with $\rho_1 = 1{,}027.5\,\text{kg/m}^3$ and $\rho_2 = 1{,}028.5\,\text{kg/m}^3$, which halves the density contrast between the layers. Equation (3.61) suggests that the thickness of the initial mixing regime will double in this situation. The reader is encouraged to verify this.

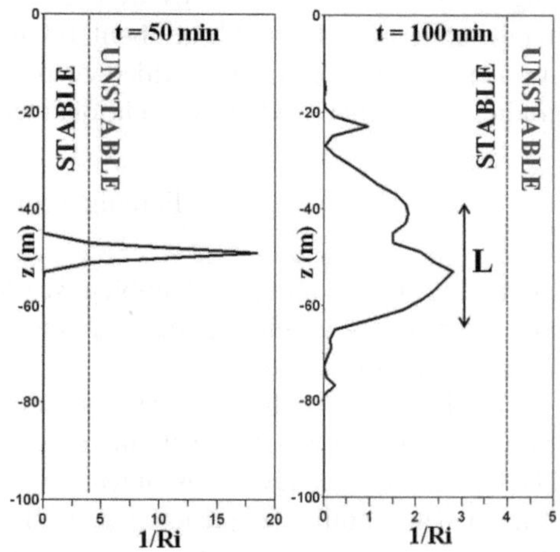

Fig. 3.26 Exercise 6. Vertical profiles of $1/Ri$ after (*left panel*) 25 min and (*left panel*) 100 min of simulation. Vertical profiles are computed from horizontal averages of density and lateral flow speed. The value $1/Ri = 4$ indicates the theoretical transition between laminar and turbulent flow

3.12 Lee Waves and the Froude Number

3.12.1 The Hydraulic Jump

Lee waves are created by flow past a topographic irregularity such as a mountain in the atmosphere or a sill in the ocean. Resultant disturbances of density interfaces in the lee of topographic obstacles are internal waves of a certain phase speed. Whether disturbances can locally grow to great amplitudes depends on the phase speed of internal waves in relation to the carrier speed of the ambient flow.

The ratio between speed of the incident flow and the phase speed of long gravity waves is called the *Froude number*, introduced by William Froude (1874). For a Froude number less than unity, wave disturbances can travel upstream against the ambient flow, a situation referred to as *subcritical*. For a Froude number is excess of unity, on the other hand, wave disturbances are carried downstream by the ambient flow. This situation is referred to as *supercritical*.

The transition between these two regimes is called the *hydraulic jump*, which implies a Froude number of unity. In this situation, wave disturbances remain trapped and can amplify. Hydraulic jumps occur when a flow of water at high, supercritical velocity discharges into a zone of lower, subcritical velocity. Lee waves of large amplitudes are the traces of such hydraulic jumps.

3.13 Exercise 7: Lee Waves

3.13.1 *Task Description*

We consider a model domain, 500 m in length and 100 m in depth, resolved by a horizontal grid spacing of $\Delta x = 5$ m and a vertical grid spacing of $\Delta z = 2$ m (Fig. 3.27). Lateral boundaries are cyclic. A small submarine sill is included of a height of 40 m and a width of about 100 m. A cosine function is used for the creation of this sill. Forcing is provided via prescription of ambient barotropic pressure-gradient force that is added as additional terms in the u-momentum equation as:

$$\text{Forcing term} = -g \frac{\partial \eta_o}{\partial x}$$

where $\partial \eta_o / \partial x$ is a prescribed ambient sea-level gradient. This forcing, which operates to gradually accelerate the flow into the positive x-direction, is switched off after 20 min of simulation.

Two different stratification scenarios are considered. The first scenario deals with a water column in which density increases initially linearly with depth (Fig. 3.27a). The background sea level is assumed to vary by 0.5 cm over the length of the model domain. The stability frequency (squared) is varied in a range between $N^2 = 0.5 \times 10^{-4}\,\text{s}^{-2}$ and $N^2 = 5 \times 10^{-4}\,\text{s}^{-2}$. The second scenario deals with two superimposed layers of different densities both 50 m in thickness (Fig. 3.27b). Each layer is slightly stratified with a stability frequency squared of $N^2 = 0.5 \times 10^{-4}\,\text{s}^{-2}$. A density interface, also called *pycnocline*, separates both layer. Density stratification across this interface corresponds to a stability frequency (squared) of $N^2 = 4.8 \times 10^{-3}\,\text{s}^{-2}$, which is stronger than any stratification considered in Scenario 1. In this scenario,

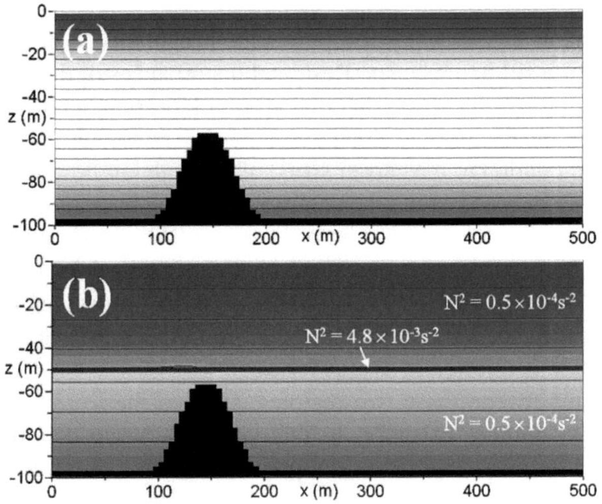

Fig. 3.27 Different initial density configurations used in Exercise 7

case studies consider variations of the barotropic forcing with the background sea level varying between 0.1 and 1.5 cm over the length of the model domain.

Density diffusion is included in both scenarios with horizontal and vertical density diffusivities being set to small uniform values of $K_h = K_z = 1 \times 10^{-4}\,\text{m}^2/\text{s}$. The total simulation time is 60 min with data outputs at one-minute intervals. The pressure accuracy for the S.O.R. iteration is set to $\epsilon = 1 \times 10^{-3}$ Pa. The time step is set to $\Delta t = 1$ s, using the rigid-lid approximation. Note that neither momentum diffusion nor boundary friction is yet included in the momentum equations.

3.13.2 Results: Continuous Density Stratification

For a weak initial stratification ($N^2 = 0.5 \times 10^{-4}\,\text{s}^{-2}$), the flow over the sill triggers a standing internal lee wave of an enormous wave height of 40 m on a wavelength of approximately 150 m (Fig. 3.28a). Near-bottom water is lifted across the sill where it becomes subject to vigorous mixing via the breaking of internal waves. This instability mechanism is responsible for the localised generation of internal waves in the ocean.

The situation is different for a stronger initial stratification ($N^2 = 5 \times 10^{-4}\,\text{s}^{-2}$) (Fig. 3.28b). In this case, the sill operates as a barrier for the flow with the consequence that flow below sill depth is almost absent. Interaction between flow at mid depth and bathymetry creates lee waves of a reduced height of 10 m. This demonstrates that, for strong density stratification, bathymetric obstacles can operate as a barrier for flows.

According to Eq. (3.55), the phase speed of internal waves in a fluid of continuous density stratification depends on the wavelength of the disturbance. In this situation, a hydraulic jump flow will always generate lee waves of a wavelength corresponding

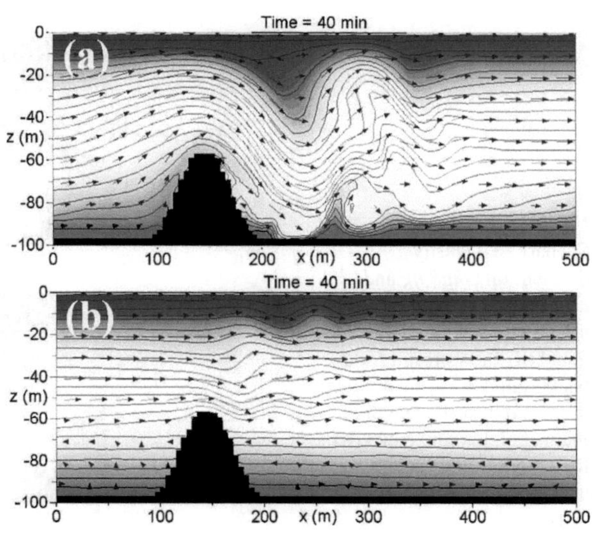

Fig. 3.28 Exercise 7. Scenario 1. Density distribution (*shading*) and flow field (*arrows*) after 40 min of simulation. Panel (**a**) shows results for an initial value of $N^2 = 0.5 \times 10^{-4}\,\text{s}^{-2}$, panel (**b**) for $N^2 = 5 \times 10^{-4}\,\text{s}^{-2}$. Flow vectors are averaged over 5 by 5 grid cells

to a Froude number of unity. We anticipate that stronger density stratification leads to smaller amplitudes of lee waves.

3.13.3 Results: Two-Layer Stratification

Existence of a pycnocline supports the creation of long interfacial waves of a phase speed of (e.g. Pond and Pickard, 1983):

$$c = \sqrt{g'h^*} \qquad (3.63)$$

where reduced gravity is $g' = \Delta\rho/\rho_o g$ with $\Delta\rho$ being the difference in layer densities, and the reduced depth scale is given by: $h^* = H_1 H_2/(H_1 + H_2)$, where H_1 and H_2, are the initial thicknesses of the upper and lower layers.

In contrast to long internal waves in a fluid of continuous density stratification, the phase speed of such waves is independent of wavelength. In this situation, the creation of a hydraulic jump exclusively depends on whether the incident flow is swift enough to become supercritical over the sill. For the density configuration of Scenario 2, the phase speed of interfacial waves is about 50 cm/s upstream the topographic obstacle, but decreases to 28 cm/s over the hill. The occurrence of a hydraulic jump and lee wave formation requires an ambient flow of speed exceeding 30 cm/s.

The choice of a sea-level gradient of 0.1 cm over the length of the model domain creates an ambient flow of maximum speed of 16 cm/s over the sill, which is not sufficient to create a hydraulic jump (not shown). The reader can easily verify this statement with own simulations. For a increased sea-level gradient of 0.5 cm, the flow speed increases to values of 35 cm/s above the sill. This is sufficient to produce interfacial waves of a height of 10 m in the lee of the sill (Fig. 3.29a).

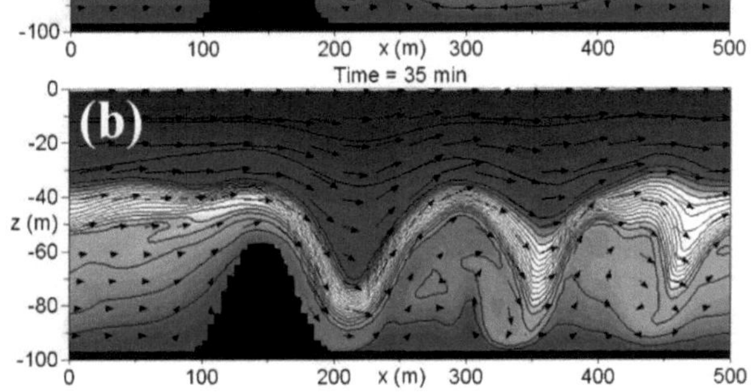

Fig. 3.29 Exercise 7. Scenario 2. Density distribution (*shading* and *lines*) and flow field (*arrows*) after 35 min of simulation. Panel (**a**) shows result for a background sea-level gradient of 0.5 cm over the length of the model domain, panel (**b**) for 1.5 cm. Flow vectors are averaged over 5 by 5 grid cells

Further enhanced forcing provided by a background sea-level gradient of 1.5 cm over the length of the domain triggers massive internal waves of a height of 40 m (Fig. 3.29b). This demonstrates that swift flows can trigger large-amplitude internal waves at density interfaces despite strong vertical density contrasts.

3.13.4 Additional Exercise for the Reader

Repeat this exercise with different shapes of topographic obstacles and/or different configurations of the initial density field. For instance, it would be interesting to explore resultant wave patterns for a system of three superimposed layers of different densities.

3.14 Oceanic Convection

3.14.1 Background

Density-driven convection is a form of instability that is triggered by unstable density stratification in fluids. Convection produces vigorous vertical mixing and, hence, aims at reestablishing a statically stable density field. The theory of density-driven convection can be traced back to early studies of Bénard (1900) and Rayleigh (1916).

Convective stirring develops when heating up a fluid from below. The heating decreases the fluid's density near the bottom and produces fluid being less dense than surrounding fluid. The associated buoyancy force triggers individual plumes of fluid moving upward. Owing to volume conservation, rising fluid has to be accompanied by sinking of ambient fluid. The result is a concert of rising and sinking fluid parcels, called *free convection*. In contrast, *forced convection* is dominated by shear-flow instabilities, discussed in Sect. 3.10.

Oceanic convection is created by a surface heat loss and associated cooling and/or by an increase in salinity owing to evaporation or new ice formation. Convection is a common process in the surface-mixed layer of the oceans. It can develop during night-time cooling. If sustained over a long time, convection in the open ocean can mix the surface water column to a depth of several hundred metres; in a few regions such as the Greenland Sea and the Labrador Sea even to great depths $> 2{,}000$ m. Marshall and Schott (1999) give a comprehensive review of the current knowledge of deep-reaching convection in the open ocean.

3.14.2 Free Convection

In the absence of a vertical shear flow, unstable density stratification only leads to convection if dynamical disturbances can grow against the effects of vertical diffusion of momentum and density. The onset of free convection can be characterised by a threshold value of the Rayleigh number, introduced by Rayleigh (1916):

$$Ra = \frac{|N^2|\, d^4}{k\nu} \tag{3.64}$$

where N is the stability frequency (which is of negative sign for unstable density stratification), d is the thickness of the unstably stratified fluid layer, k is molecular diffusivity of density, and ν is kinematic molecular viscosity. Molecular diffusion is associated with random motions of molecules (Brownian motion). For a turbulent background fluid, some scientists use turbulent equivalents in Eq. (3.64).

3.14.3 The Flux-Rayleigh Number

The appearance of oceanic convection is the result of a surface buoyancy flux created by air-sea fluxes of heat and freshwater (changing salinity). A linearised equation of state is sufficient to derive surface buoyancy fluxes as a function of these heat and freshwater fluxes. This equation reads:

$$\rho(T, S) = \rho_o\,[1 - \alpha(T - T_o) + \beta(S - S_o)] \tag{3.65}$$

where the reference density ρ_o corresponds to a reference temperature T_o and a reference salinity S_o. The parameter α is the *thermal expansion* coefficient and β is an equivalent coefficient, sometimes called *saline contraction* coefficient, for salinity-related variations in density. The surface buoyancy flux B is defined indirectly via associated density changes in the surface mixed layer of thickness h that follow:

$$\frac{d\rho_{\text{mix}}}{dt} = \frac{\rho_o}{gh} B \tag{3.66}$$

With use of the equation of state (Eq. 3.65) and thermodynamic relations between surface temperature changes with surface heat fluxes and salinity changes with surface freshwater fluxes, the surface buoyancy flux can be calculated from:

$$B = g\left[\frac{\alpha}{\rho_o C_P} Q + \beta(E - P) S_{\text{mix}}\right]$$

where C_P is the heat capacity of water, Q is the surface heat flux, S_{mix} is salinity of the surface mixed layer, and $(E - P)$ is evaporation minus precipitation (units are m/s). Gill (1982) presents tables of α, β, and C_P for seawater as functions of ambient temperature and salinity.

Free convection develops if a certain threshold value of the *flux-Rayleigh number* is exceeded, whereby this number is defined by (Marshall and Schott, 1999):

$$Ra_f = \frac{Bd^4}{k^2\nu}$$

Considerations of Rayleigh numbers are useful for studies of the onset of convection in laboratory settings and computer-based simulations. Values of critical flux-Rayleigh numbers in the real situation, however, are not known with any certainty.

3.14.4 Aspect Ratio of Convection Cells

The distance of adjacent convective plumes is typically twice their penetration depth (Turner, 1973), which can be expressed in terms of a geometric ratio called *aspect ratio* of convection cells (Fig. 3.30). This knowledge is useful for adequate choices of grid spacings and model-domain size. Recall that the length scale of a process should be spatially resolved by at least 10 grid points and that the size of the model domain should exceed this length scale several times. A penetration depth of 100 m, for instance, suggests the use of horizontal and vertical grid spacings of <10 m, while the model domain should have a width of approximately 1 km, such that a sufficient number of convection cells can develop inside the domain.

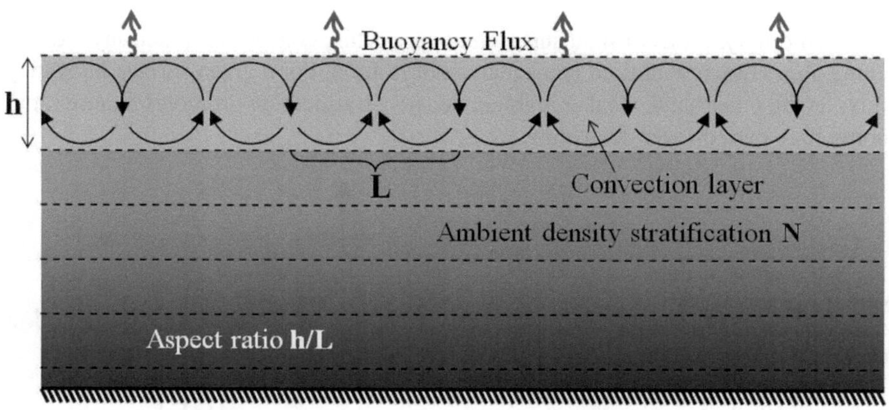

Fig. 3.30 Illustration of convection in the surface mixed layer of the ocean

3.14.5 Convective Mixed-Layer Deepening

Consider a water column in which density initially increases linearly with depth according to:

$$\rho(z) = \rho_o \left(1 + \frac{N^2}{g}|z|\right)$$

where ρ_o is the initial surface density, and the stability frequency is given by Eq. (1.6). After convective mixing to a certain depth h (Fig. 3.31), the density throughout the surface mixed layer equals that found at depth h; that is,

$$\rho_{mix} = \rho_o\left(1 + \frac{N^2}{g}h\right)$$

The temporal derivative of the latter equation gives:

$$\frac{d\rho_{mix}}{dt} = \rho_o \frac{N^2}{g}\frac{dh}{dt}$$

Using Eq. (3.66) gives the relation (Turner, 1973):

$$h\frac{dh}{dt} = 0.5\frac{dh^2}{dt} = \frac{B}{N^2} \tag{3.67}$$

which has the solution:

$$h(t) = \sqrt{\frac{2\,B}{N^2}} \cdot t$$

Accordingly, for a constant surface buoyancy flux, convective mixed-layer deepening slows down with time in a square-root fashion. For a given surface buoyancy flux, on the other hand, weaker ambient density stratification promotes a more rapid convective mixed-layer deepening.

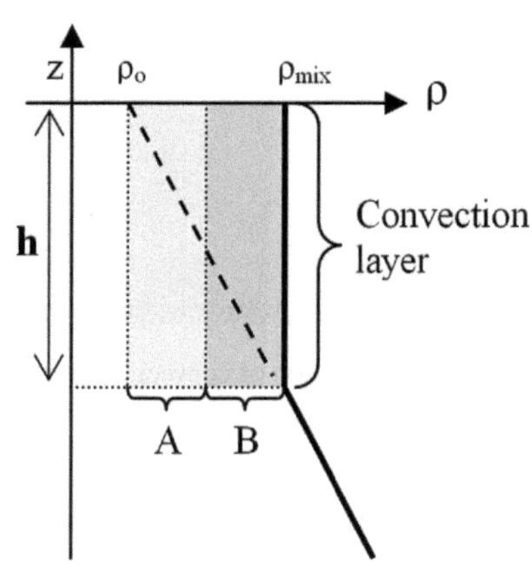

Fig. 3.31 Contributions to density increases in the surface mixed layer subject to convective mixing. Half the total surface density increase (symbolised by **A**) is provided by mixing of the initial ambient density gradient over the depth h, the other half (symbolised by **B**) is supplied by the surface buoyancy flux

3.15 Exercise 8: Free Convection

3.15.1 Aim

The aim of this exercise is to simulate the free convection process in the ocean with the vertical ocean-slice model.

3.15.2 Task Description

Consider an ocean slice of 100 m in depth and 1,000 m in length (Fig. 3.32), resolved by equidistant grid spacings of $\Delta x = \Delta z = 5$ m. Cyclic boundary conditions are used at the lateral boundaries. A linear version of the equation of state is used (Eq. 3.65), whereby salinity effects are ignored. The thermal expansion coefficient is set to a value of $\alpha = 2.5 \times 10^{-4}\,\mathrm{K}^{-1}$. Initially, the water column is at rest and stably stratified in temperature with a stability frequency of $N = 10^{-3}\,\mathrm{s}^{-1}$. Random density fluctuations with maximum values of $10^{-4}\,\mathrm{kg/m^3}$ are added to the density field using the random-number generator of previous exercises.

A uniform heat loss of $Q = 600\,\mathrm{W/m^2}$ is prescribed at the sea surface. In the absence of other processes (such as downward diffusion) this heat loss would cool the uppermost grid cell at a rate of:

$$\frac{\partial T_s}{\partial t} = -\frac{Q}{\rho_o C_P \Delta z}$$

where C_P is the heat capacity of seawater which has a value of around $C_P = 4,000\,\mathrm{J\,kg^{-3}\,K^{-1}}$. The corresponding temperature decrease would be 1.2°C per day for a water basin of 5-m depth, which is the vertical grid space used. On the basis of Eq. (3.65), surface density in this depth-confined basin would change according to:

$$\frac{\partial \rho_s}{\partial t} = \frac{\alpha Q}{C_P \Delta z}$$

where the sign of heat flux is defined such that positive values incur a density increase. In addition to isotropic density diffusion with small values of diffusivity of $K_h = K_z = 10^{-4}\,\mathrm{m^2/s}$, isotropic diffusion of momentum is added to the

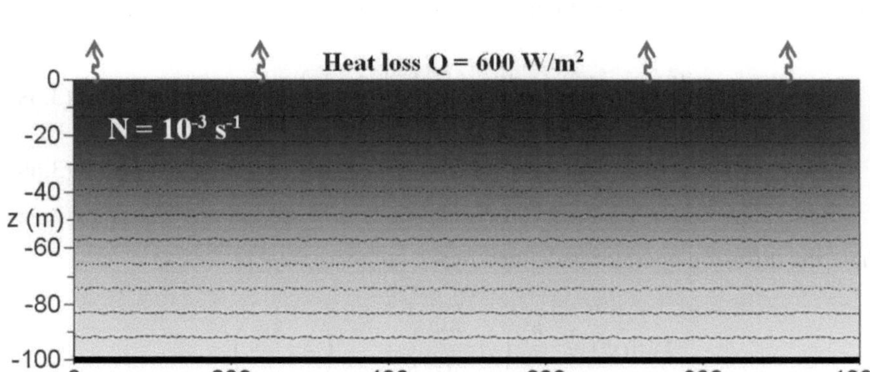

Fig. 3.32 Initial configuration for Exercise 8. *Shading* and *lines* show the initially stable density stratification of the water column

code with a viscosity of $A_h = A_z = 10^{-4}\,\mathrm{m^2/s}$, which is close to molecular values. This also gives us the opportunity to include a quadratic bottom-friction law, reading:

$$\frac{\tau_x^{\mathrm{bot}}}{\rho_o} = \left(A_z \frac{\partial u}{\partial z}\right)_{z=-h} = r\ u_b\ |u_b|$$

where u_b is the horizontal flow adjacent to the sea floor. The bottom-drag coefficient is set to $r = 0.001$. The total simulation time is 6 hrs with data outputs at every 3 min. The time step is set to $\Delta t = 1$ s. The pressure accuracy of the S.O.R. iteration is set to $\epsilon = 1 \times 10^{-3}$ Pa. This model application uses the rigid-lid approximation (see Sect. 3.7).

3.15.3 A Trick to Avoid Substantial Round-off Errors

Owing to small numerical time steps, density increments caused by surface cooling are very small ($10^{-8}\,\mathrm{kg/m^3}$) compared with the ambient density ($1{,}028\,\mathrm{kg/m^3}$). When using true seawater density in the code, these minute density increments tend to get lost due to round-off errors. The consequence is that you cannot simulate the convection process when using true density. The trick to avoid this is to rather use relative density $(\rho - \rho_s)$, where ρ_s is the initial surface value, as a variable in the code, which significantly increases the accuracy of the prediction.

3.15.4 Inclusion of Momentum Diffusion and Bottom Friction

The momentum equations including diffusion terms can be written as:

$$\frac{\partial u}{\partial t} + u\frac{\partial u}{\partial x} + w\frac{\partial u}{\partial z} = -\frac{1}{\rho_o}\frac{\partial (p+q)}{\partial x} + \mathrm{Diff}(u) \tag{3.68}$$

$$\frac{\partial w}{\partial t} + u\frac{\partial w}{\partial x} + w\frac{\partial w}{\partial z} = -\frac{1}{\rho_o}\frac{\partial q}{\partial z} + \mathrm{Diff}(w) \tag{3.69}$$

where the diffusion operator is defined by:

$$\mathrm{Diff}(\psi) = \frac{\partial}{\partial x}\left(A_h \frac{\partial \psi}{\partial x}\right) + \frac{\partial}{\partial z}\left(A_z \frac{\partial \psi}{\partial z}\right)$$

where the symbol ψ stands for either u or w. This formulation includes spatially variable values of eddy viscosities A_h and A_z, which is considered in later exercises.

Horizontal and vertical diffusion of u can be formulated in finite-difference form as:

$$\frac{\partial}{\partial x}\left(A_h \frac{\partial u}{\partial x}\right) = \frac{A_x^e(u_{i,k+1} - u_{i,k}) - A_h^w(u_{i,k} - u_{i,k-1})}{(\Delta x)^2} \tag{3.70}$$

$$\frac{\partial}{\partial z}\left(A_z \frac{\partial u}{\partial z}\right) = \frac{A_z^+(u_{i-1,k} - u_{i,k}) - A_z^-(u_{i,k} - u_{i+1,k})}{(\Delta z)^2} \tag{3.71}$$

where interpolated values of eddy viscosities are given by:

$$A_h^e = A_{h,i,k+1}$$
$$A_h^w = A_{h,i,k}$$
$$A_z^+ = 0.25\left(A_{z,i,k} + A_{z,i,k+1} + A_{z,i-1,k} + A_{z,i-1,k+1}\right)$$
$$A_z^- = 0.25\left(A_{z,i,k} + A_{z,i,k+1} + A_{z,i+1,k} + A_{z,i+1,k+1}\right)$$

Horizontal and vertical diffusion of w is calculated in a similar fashion from:

$$\frac{\partial}{\partial x}\left(A_h \frac{\partial w}{\partial x}\right) = \frac{A_h^e(w_{i,k+1} - w_{i,k}) - A_h^w(w_{i,k} - w_{i,k-1})}{(\Delta x)^2} \tag{3.72}$$

$$\frac{\partial}{\partial z}\left(A_z \frac{\partial w}{\partial z}\right) = \frac{A_z^+(w_{i-1,k} - w_{i,k}) - A_z^-(w_{i,k} - w_{i+1,k})}{(\Delta z)^2} \tag{3.73}$$

where:

$$A_h^e = 0.25\left(A_{h,i,k} + A_{h,i,k+1} + A_{h,i-1,k} + A_{h,i-1,k+1}\right)$$
$$A_z^w = 0.25\left(A_{h,i,k} + A_{h,i,k-1} + A_{h,i-1,k} + A_{h,i-1,k-1}\right)$$
$$A_z^+ = A_{z,i-1,k}$$
$$A_z^- = A_{z,i,k}$$

These diffusion terms, multiplied with the numerical time step, are added as additional components to the u^* and w^* arrays in Eq. (3.45) and (3.46). Bottom friction is implemented in the bottom-nearest grid cell via a quadratic bottom-friction law formulated as:

$$A_z^- \frac{u_{i,k} - u_{i+1,k}}{\Delta z} = r\, u_{i,k} \left|u_{i,k}\right|$$

The above finite-difference forms of the diffusion terms are associated the stability condition:

$$\Delta t \le \min\left\{\frac{(\Delta z)^2}{\max(A_z)}, \frac{(\Delta x)^2}{\max(A_h)}\right\}$$

In comparison, the CFL stability criteria (Eqs. 3.35 and 3.43) are often (but not always!) associated with relatively shorter time steps.

3.15.5 Results

Surface cooling creates unstable density stratification in vicinity of the sea surface (Fig. 3.33). The degree of unstable stratification increases over time to a point when the critical Rayleigh number is exceeded and convection follows. This onset of convection takes place in this experiment after approximately 40 min of simulation. Convection cells appear near the surface on a fairly small spatial scale set by the grid spacing prescribed. Convection cells comprise individual sinking plumes and compensating upward flows. Individual plumes attain vertical speeds of up to 12 cm/s.

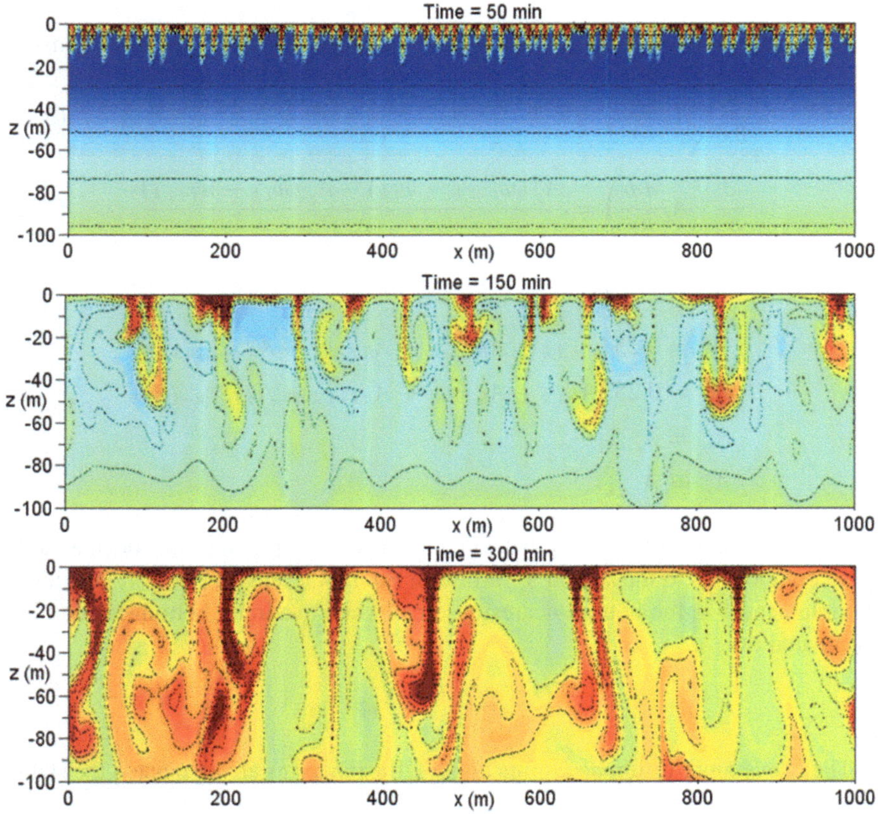

Fig. 3.33 Exercise 8. Distributions of relative density ($\rho - \rho_s$) (*color shading* and *contours*) at selected times of the simulation. Shown are values of relative density in a range between 0 and 0.02 kg/m^3

Given that the ambient water column is initially stably stratified, the penetration depth of the convective plumes gradually increases with time. The reader is encouraged to compare the simulated rate of mixed-layer deepening with that predicted by the solution to Eq. (3.67). The spacing between plumes increases as the penetration reaches greater depths. The predicted aspect ratio of convection cells of 1.5–2 is in agreement with theory. Convective plumes reach the sea floor at 100-m depth after about 3 hrs of applied surface heat loss. The final result of convection is an almost well-mixed water column.

3.15.6 Additional Exercise for the Reader

Repeat the convection exercise with a fluid of two superimposed layers of different densities. Do convective plumes induce internal waves at the density interface? The reader is also encouraged to repeat this exercise with a stronger initial density stratification.

3.16 Exercise 9: Convective Entrainment

3.16.1 How It Works

Convection produces vigorous mixing in upper parts of an otherwise stratified water column. As the convection layer deepens, convective plumes mix deeper water into this layer – a process which is called *entrainment*. Convective entrainment of heat plays an important role in polar oceans, for it can delay or even prevent sea ice formation, which would occur if the surface water were cooled down to temperatures below the freezing-point temperature.

3.16.2 Entrainment Velocity

For an ocean of initially stable density stratification, Eq. (3.67) implies that the thickness of the convective mixed layer h increases at a rate of:

$$\frac{dh}{dt} = \frac{B}{N^2 h}$$

This mixed-layer deepening can be directly taken as a *vertical entrainment velocity*. This velocity determines how much heat, salt and other water properties become mixed into the convection layer.

3.16.3 Task Description

For illustration of the convective entrainment process, we repeat Exercise 8 with addition of both Eulerian tracer concentration and non-buoyant Lagrangian floats. To this end, an advection-diffusion equation for tracer concentration C is added to the code reading:

$$\frac{\partial C}{\partial t} + \text{Adv}(C) = \text{Diff}(C)$$

where the same eddy diffusivities are used as in the density conservation equation. Concentrations are initialised with values of unity in a 25-m thick near-bottom layer and zero values elsewhere. In conjunction with this, 3,000 Lagrangian floats are initially distributed at random locations within 25 m from the sea floor. Changes of float locations (x^*, z^*) are calculated from simple displacement equations:

$$\frac{dx^*}{dt} = u \quad \text{and} \quad \frac{dz^*}{dt} = w$$

where (u, w) is the velocity predicted by model in vicinity of the float. Instead of accurate interpolation of velocity to the precise location of a float, it is sufficient for the purpose of this exercise to use the velocity in the grid cell surrounding a float as a proxy. Hereby, velocity components are interpolated to pressure grid points and the grid cell containing a float is defined within $\pm 0.5\Delta x$ distance horizontally and $\pm 0.5\Delta z$ distance vertically with respect to this grid point. Special treatment is required to avoid stranding of floats in dry grid cells. Section 5.8 describes a more accurate method.

The total simulation time is 6 hrs with data outputs at every 2.5 min of the simulation. The time step is set to $\Delta t = 1$ s using the rigid-lid approximation. Note that the model gives similar results with inclusion of a free surface. This, however, requires a shorter time step of $\Delta t = 0.1$ s, leading to a tenfold increase of the total simulation time.

3.16.4 Results

It takes about 2 hrs of applied surface buoyancy flux until convective plumes reach into the bottom layer to entrain near-bottom water into the convection layer (Fig. 3.34). Again, we can see that the product of convection is an almost perfectly mixed water column. As expected, Lagrangian floats are also entrained into the convection layer (Fig. 3.35). We can imagine that these floats represent small sediment particles being mixed into the water column.

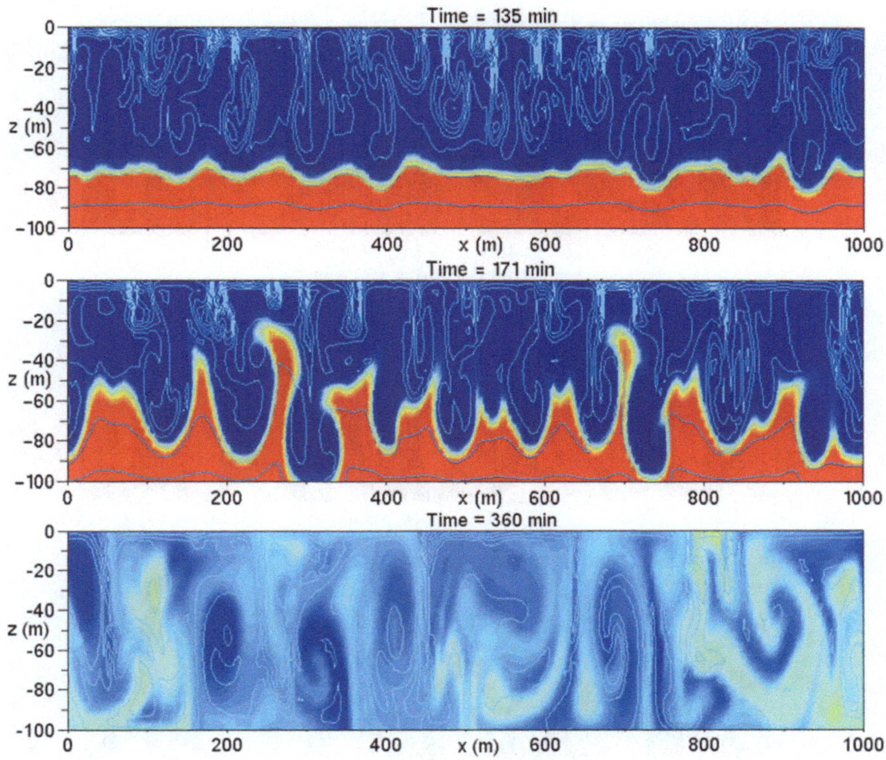

Fig. 3.34 Exercise 9. Distributions of Eulerian tracer concentration (*color shading*) at selected times of the simulation. *Shading* refers to a range in concentration between 0 and 1. *Lines* are density contours

3.16.5 Additional Exercises for the Reader

Produce a Scilab script that produces a graph of the pathways, called *trajectories*, of selected floats. What is the greatest horizontal distance travelled by a float?

3.17 Exercise 10: Slope Convection near the Shore

3.17.1 Background

Under the same heat loss, a smaller volume of fluid cools down at a faster rate than a larger volume. Therefore, we expect that convection produces the densest water in the shallowest regions of a coastal ocean. In the presence of variable bathymetry, a uniform surface buoyancy flux will consequently create lateral density gradients and, hence, lateral pressure gradients in the ocean. A buoyancy-driven exchange

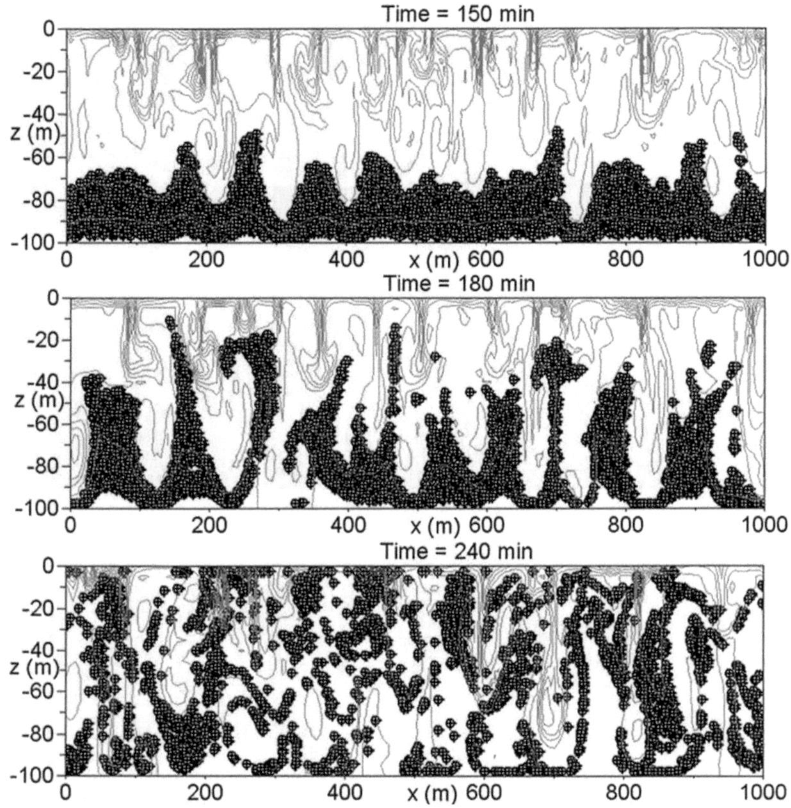

Fig. 3.35 Exercise 9. Distributions of 3,000 Lagrangian floats at selected times of the simulation. *Lines* are density contours

circulation between shallow and deeper portions of the ocean follows. The aim of this exercise is to simulate this density-driven overturning circulation using the non-hydrostatic convection model in a domain of variable bathymetry.

3.17.2 Implementation of Bottom Friction on a Sloping Terrain

The stair-step nature of z-coordinate systems requires special treatment for the calculation of bed-shear stresses. This needs to include all wet u-velocity grid points situated directly above the sea floor, but also those that sit directly above the corner of a step. Figure 3.36 gives an example.

3.17.3 Task Description

We consider a vertical ocean slice of 1 km in length and variable bathymetry, resolved by equidistant grid spacings of $\Delta x = \Delta z = 5\,\text{m}$ (Fig. 3.37). Lateral

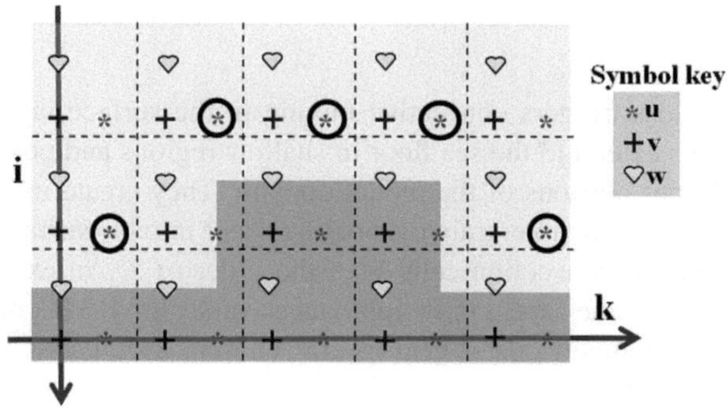

Fig. 3.36 *Circles* indicate u-velocity grid points used for calculation of bed shear stresses

boundaries are closed. The bathymetry consists of a sloping transition zone that connects a shallow region (20 m) with a deeper region (100 m). Initially, the ocean is at rest and consists of two superimposed layers of different densities. The density interface is located at a depth of 50 m and is characterised by a stability frequency of $N = 7.6 \times 10^{-3}\,\mathrm{s}^{-1}$. Each of the two layers is slightly stratified in density with a stability frequency of $N = 1 \times 10^{-3}\,\mathrm{s}^{-1}$. Random density fluctuations with maximum values of $10^{-4}\,\mathrm{kg/m^3}$ are added to the density field using the random-number generator of previous exercises.

Eddy diffusivities, eddy viscosities and the bottom-drag coefficient are the same as in Exercise 8. The time step is set to $\Delta t = 2$ s using the rigid-lid approximation. The pressure accuracy for the S.O.R. iteration is set to $\epsilon = 1 \times 10^{-3}$ Pa. The total simulation time is 12 hrs with data outputs at 6-min intervals. A uniform heat loss of $Q = 600\,\mathrm{W/m^2}$ is prescribed at the sea surface for the first 6 hrs of simulation. The heat loss is set to zero for the rest of the simulation to explore the density-driven adjustment process that follows. Note that such a forcing does rarely ever occur in nature. It is applied her for demonstration purposes only. Eulerian tracer concentration is added to the water column with initial values of unity in the upper 20 m of the water column and zero values elsewhere.

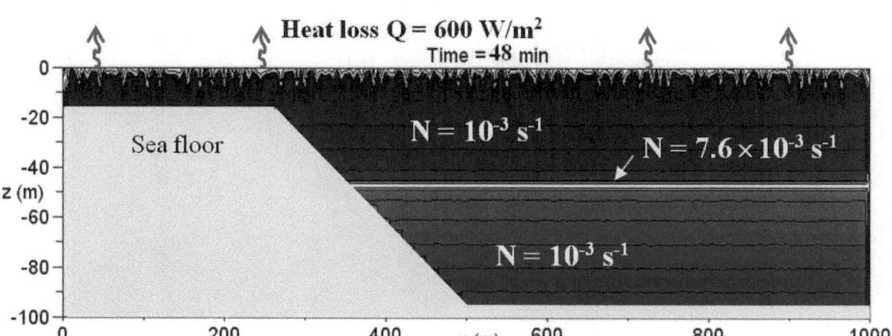

Fig. 3.37 Exercise 10. Density distribution (*shading* and *lines*) after 48 min of the simulation

3.17.4 Results

The surface heat loss triggers convective mixing in the surface layer (Fig. 3.38). Convective plumes extend to the sea floor in shallow regions and penetrate into the pycnocline in deeper portions of the model domain. They create internal waves at the pycnocline that operate to entrain pycnocline water into the surface mixed layer.

The aspect ratio of convection cells is of the order of 2. An exception are the shallow regions where the aspect ratio approaches values of 4. This bias is presumably caused by the relatively coarse grid spacing chosen, which does not adequately resolve the length scale of convection cells in this region.

Owing to limited penetration, the density gain in shallow water exceeds that in deeper water, so that a density contrast between shallower and deeper regions develops over time. With the vanishing surface heat flux, the dense water formed in the shallow regions forms of a gravity current flowing downward on the slope. This triggers a return flow in upper parts of the water column, which is a requirement of volume conservation, driven by a lateral pressure gradient owing to an inclined sea surface.

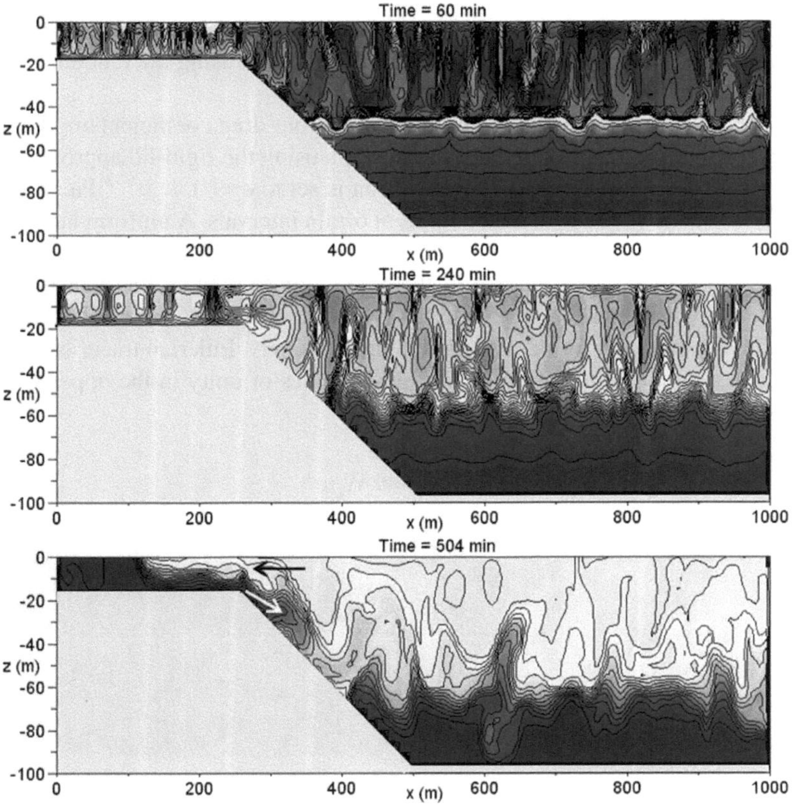

Fig. 3.38 Exercise 10. Snapshots of density distribution (*shading* and *contours*) at selected times of the simulation

The evolution of Eulerian tracer concentration reveals that the density-driven current leaves the slope and spreads horizontally at a depth of approximately 50 m, where it encounters ambient water of equal density (Fig. 3.39). Hence, convection in shallow water can operate to ventilate sub-surface water in adjacent deeper layers of the water column. Interestingly, the laterally injected water layer, being 10 m thick, forms a large-amplitude undulation as it becomes subject to internal wave motions of the pycnocline.

It should be noted that bottom-arrested flows on a step-structured sea floor of z-coordinate models are substantially biased because their zigzag pathway is much longer than that of real bottom-parallel flows. Another artifact is that, while descending on the bathymetric stairs, the head of a plume has to spread laterally first before it can sink to the next deeper level. This induces transient unstable density stratification, which in reality does not exist, that may bias the dynamics via artificial convective adjustments.

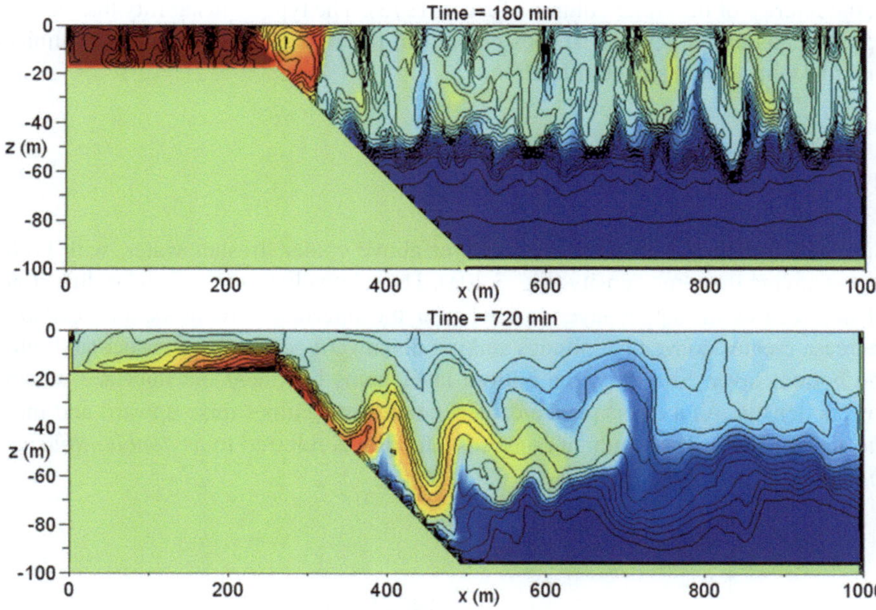

Fig. 3.39 Exercise 10. Snapshots of distribution of Eulerian tracer concentration (*color shading*) at selected times of the simulation. *Lines* are density contours

3.17.5 Additional Exercise for the Reader

Extend the simulation time of this exercise to 24 hrs and apply the surface heat flux over the initial 12 hrs. Note that, for simplicity, rotational effects owing to the Coriolis force, that could modify the dynamics, are still neglected. How deep does the resultant dense water flow penetrate into the ambient water column? The reader is also encouraged to add the Lagrangian float prediction scheme to the model code as another means to track movement of the density-driven plume.

3.18 Double Diffusion

3.18.1 Background

Molecules perform random motions, called *Brownian motion* (Brown, 1866), in the absence of turbulence. The result of this motion is a slow but continuous *molecular diffusion* which operates to smooth curvature in spatial distributions of a property. The rate of molecular diffusion of heat is about 100 times that for salt. Hence, molecular diffusion in a water column uniform in density but gradients in temperature and salinity gives rise to local density variations that can trigger the onset of convection in the interior of the water column (Turner, 1973). The type of instability that develops depends on the specific shapes of vertical profiles of temperature and salinity that make up the density stratification.

3.18.2 Double-Diffusive Instability

Suppose there is a warm and saltier layer above cooler, fresher water, with both layers having the same density (Fig. 3.40a). The relatively faster molecular diffusion of heat causes the upper layer to cool near the interface. The associated density increase creates convective plumes sinking downward across the interface and into the bottom layer. Similarly, the bottom layer gains heat near the interface which lowers density. As a consequence of this, convective plumes raise upward and into the surface layer. The resultant convective mixing is referred to as *double-diffusive instability* or *salt fingering*.

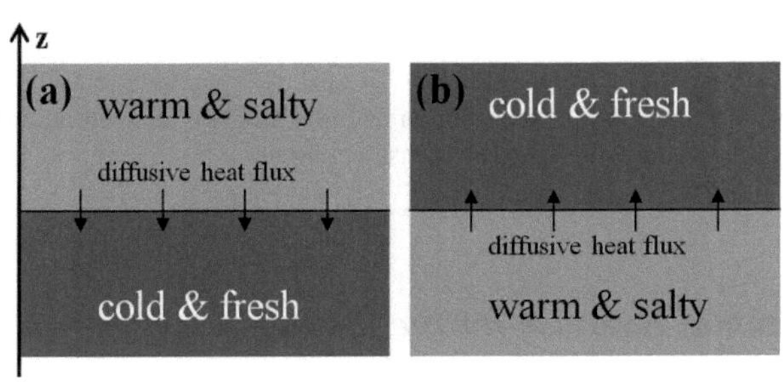

Fig. 3.40 Different stratification scenarios of a water column uniform in density leading to distinct double-diffusive processes. Scenario (**a**) causes double-diffusive mixing across the interface. Scenario (**b**) leads to isolated convective mixing in each layer and a density contrast develops between the layers

3.18.3 Double-Diffusive Layering

Suppose there is a cooler, fresher layer above warmer, saltier water with the same density (Fig. 3.40b). In this case, molecular diffusion of heat warms the surface layer from below. As a result of this, convection develops in the top layer. Similarly, cooling from above triggers convection in the bottom layer. In contrast to the double-diffusive instability, there is only little mixing across the interface. Moreover, the molecular heat flux at the interface leads to the formation of a density difference between the layers. The final result is a layer structure with a sharp density contrast across the interface. This situation is referred to as *double-diffusive layering*.

3.18.4 The Gradient Ratio and the Turner Angle

A *gradient ratio* can be defined by:

$$R_\rho = \frac{\alpha \partial T/\partial z}{\beta \partial S/\partial z} \tag{3.74}$$

where α is the thermal expansion coefficient, and β is the salinity coefficient in the equation of state (Eq. 3.65). This ratio gives the relative contributions of thermal gradients and saline gradients to the density stratification. For a two-layer fluid, the latter relation can be written as:

$$R_\rho = \frac{\alpha \Delta T}{\beta \Delta S} \tag{3.75}$$

where ΔT and ΔS are temperature and salinity contrasts between the layers. Since the gradient ratio can attain infinite values, it is more convenient to use the so-called *Turner angle*, which is defined by (Ruddick, 1983):

$$Tu = \arctan\left(\frac{R_\rho - 1}{R_\rho + 1}\right) \tag{3.76}$$

The Turner angle is used for the classification of different dynamic mixing regimes that can develop in the ocean (Fig. 3.41). Turner angles $|Tu| > 90°$ characterise unstable density stratification supporting the onset of thermohaline free convection (see Sect. 3.14). Turner angles of $\pm 90°$ characterise situations of vanishing density gradients, whereas Turner angles in a range between $\pm 45°$ and $\pm 90°$ can lead to either double-diffusive instability or double-diffusive layering. Turner angles less than $\pm 45°$ correspond to a stably stratified configuration that does not support either of these processes.

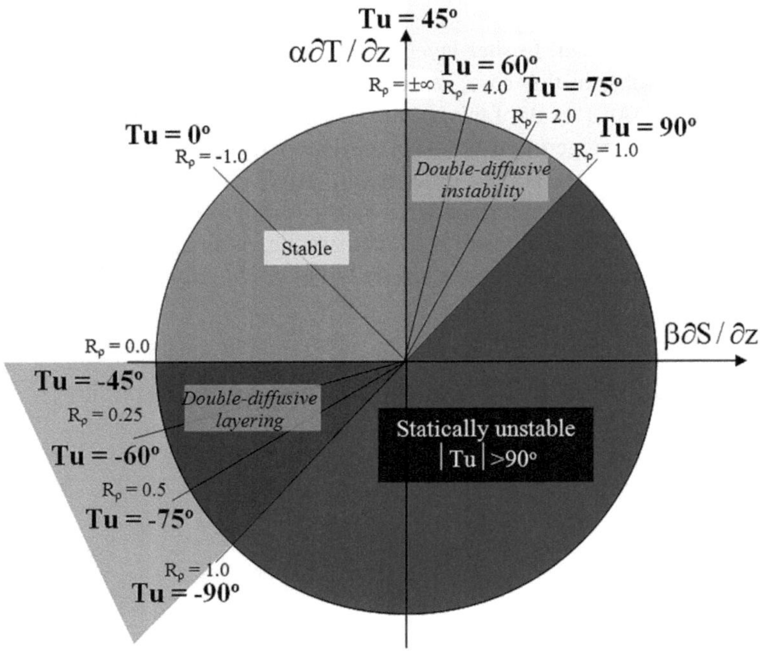

Fig. 3.41 Dynamical regimes for different gradient ratios and Turner angles

3.19 Exercise 11: Double-Diffusive Instability

3.19.1 Aim

The aim of this exercise is to explore the double-diffusive instability mechanism with application of the nonhydrostatic ocean-slice model.

3.19.2 Task Description

We consider a model domain, 200 m in length and 20 m deep, resolved by equidistant grid spacings of $\Delta x = \Delta z = 1$ m. Lateral boundaries are cyclic. Advection-diffusion equations for temperature and salinity together with a linearised equation of state are added to the model code. Diffusivity of heat is set to a uniform value of 10^{-4} m^2/s and diffusivity of salt is chosen as 10^{-6} m^2/s. These values, exceeding true values by a factor of 1,000, are deliberately chosen to speed up the simulation.

Rather than full values, we use temperature anomalies T' and salinity anomalies S' with reference to constant background values as field variables. Accordingly, the linearised equation of state can be written as:

$$\rho' = \rho - \rho_o = \rho_o(\beta S' - \alpha T') \tag{3.77}$$

where ρ' is density anomaly, the thermal expansion coefficient is set to $\alpha = 2.5 \times 10^{-4}\,\mathrm{K}^{-1}$, and the salinity coefficient is set to $\beta = 8 \times 10^{-4}$. The advection-diffusion equations for temperature and salinity anomalies read:

$$\frac{\partial T'}{\partial t} + \mathrm{Adv}(T') = K_T \left(\frac{\partial^2 T'}{\partial x^2} + \frac{\partial^2 T'}{\partial z^2} \right)$$

$$\frac{\partial S'}{\partial t} + \mathrm{Adv}(S') = K_S \left(\frac{\partial^2 S'}{\partial x^2} + \frac{\partial^2 S'}{\partial z^2} \right)$$

where K_T and K_S are molecular diffusivities. Note that temperature and salinity values are calculated at the same grid points as density. The advection-diffusion equations for temperature and salinity have the same form as the density conservation equation, used in previous exercises, except for different values of diffusivities. Inclusion of these equations in the simulation code is therefore a simple copy-and-paste and renaming task.

The initial density field consists of two superimposed layers of the same density and a thickness of 10 m each. The top layer is $T' = 10°$C warmer compared with the bottom layer. Small random disturbances in temperature of $10^{-4\circ}$C in magnitude are added. The salinity excess is calculated from Eq. (3.77) such that the initial density field is spatially uniform. The initial gradient ratio is $R_\rho = 1$. The initial Turner angle is $Tu = +90°$ (see Fig. 3.40). Eddy viscosity is set to a uniform value of $A_h = A_z = 10^{-4}\,\mathrm{m}^2/\mathrm{s}$. The bottom-friction parameter is set to $r = 0.001$.

A total of 3,000 Lagrangian non-buoyant floats are randomly distributed across the surface layer for visualisation of the double-diffusive instability process. The total simulation time is 1 hr with data outputs at every min. The time step is set to $\Delta t = 1$ s using the rigid-lid approximation. Pressure accuracy of the S.O.R. iteration is set to $\epsilon = 1 \times 10^{-3}$ Pa. In order to monitor the change in thermal and saline contrasts between the layers, we also calculate layer-averaged values of temperature and salinity in each layer to determine the trends of $\alpha \Delta T$ and $\beta \Delta S$.

3.19.3 Results

Diffusion of heat creates a thin layer of unstable density stratification in vicinity of the interface (Figs. 3.42 and 3.43). Small-scale dynamic instabilities appear in vicinity of the interface within the first 6 min of simulation triggering convective mixing of fluid parcels across the density interface. The size of convection cells gradually increases with time. Individual convective plumes attain maximum vertical speeds of 5–6 cm/s. Density anomalies inherent with the double-diffusive instability attain magnitudes of ± 0.1–$0.2\,\mathrm{kg/m}^3$. Evolution of the locations of Lagrangian floats (not shown) visualises the tendency of mixing over the entire water column.

The double-diffusion process is independent of direction and also operates in vicinity of sharp lateral temperature gradients across the "skin" of individual plumes. Hence, forcing involved in the double-diffusive instability mechanism is of complex

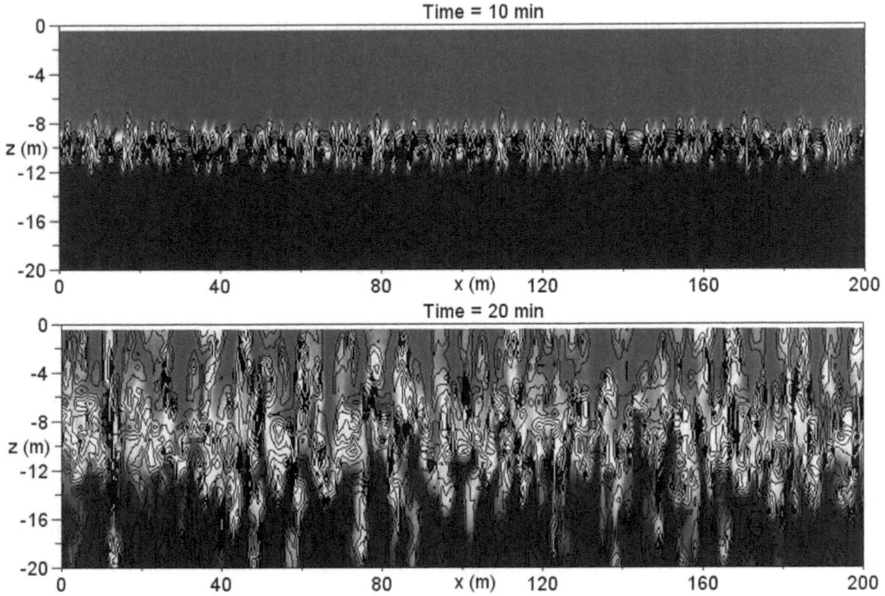

Fig. 3.42 Exercise 11. Snapshots of the temperature distribution (*shading*) at selected times of the simulation. Lines are density contours

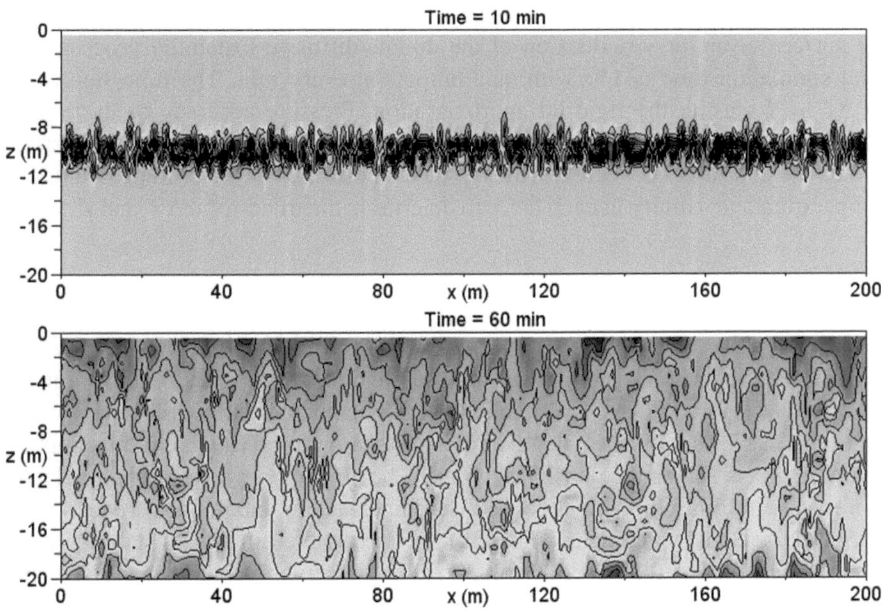

Fig. 3.43 Exercise 11. Snapshots of the density distribution (*shading* and *contours*) at selected times of the simulation

three-dimensional nature. The result of the double-diffusive instability is a slightly stratified water column characterised by a stability frequency of $N = 4\times10^{-3}$ s^{-1}. This weak residual density stratification is caused by the initial outburst of convective plumes moving density anomalies away from the initial density interface. In due course of the double-diffusive instability process, the Turner angle remains close to a value of 90° while the temperature and salinity contrasts between the layers gradually become weaker (Fig. 3.44).

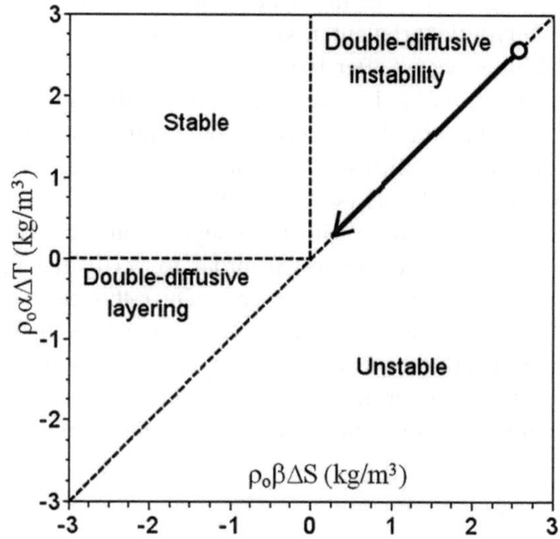

Fig. 3.44 Exercise 11. Evolution of temperature and salinity contrasts between the layers (*arrow*) expressed in equivalent density units. The *open circle* denotes the initial value. The *arrow head* points to the final value after 60 min of simulation

3.20 Exercise 12: Double-Diffusive Layering

3.20.1 Aim

The aim of this exercise is to explore the double-diffusive layering process in a water column which is initially uniform in density, but displays sharp vertical contrasts in temperature and salinity.

3.20.2 Task Description

This exercise is a repeat of Exercise 11 except for a vertical swap of the layers. The warm, saline layer is located this time below a layer of colder, fresher surface water. The temperature difference between the layers is 10°C and the salinity contrast is chosen such that density is spatially uniform. The initial gradient ratio is again $R_\rho = 1$, but the initial Turner angle is $Tu = -90°$ (see Fig. 3.41). The total simulation time is 1 hr with data outputs at 1-min intervals.

3.20.3 Results

Molecular heat exchange across the interface creates convection in each layer separately (Fig. 3.45). Only little turbulent mixing occurs across the interface. The onset of convective instabilities occurs after 10 min of simulation. The aspect ratio of convection cells is between 1 and 2. In several instances, individual convective plumes manage to plunge into the other layer causing some entrainment of fluid across the density interface. The double-diffusive layering process leads to the creation of a density contrast between the layers (see Fig. 3.45).

The author repeated this exercise with a prolonged total simulation time of 24 hrs (which took several hours to complete) in order to verify the final state resulting from the double-diffusive layering process. Over this time span, the Turner angle increased from $-90°$ to $-78°$ (Fig. 3.46), which is in stark contrast to the double-diffusive instability process (see Fig. 3.44). This is again evidence of the formation of a density interface between the layers, but also shows the tendency of slow erosion of temperature differences while a certain salinity contrast remains.

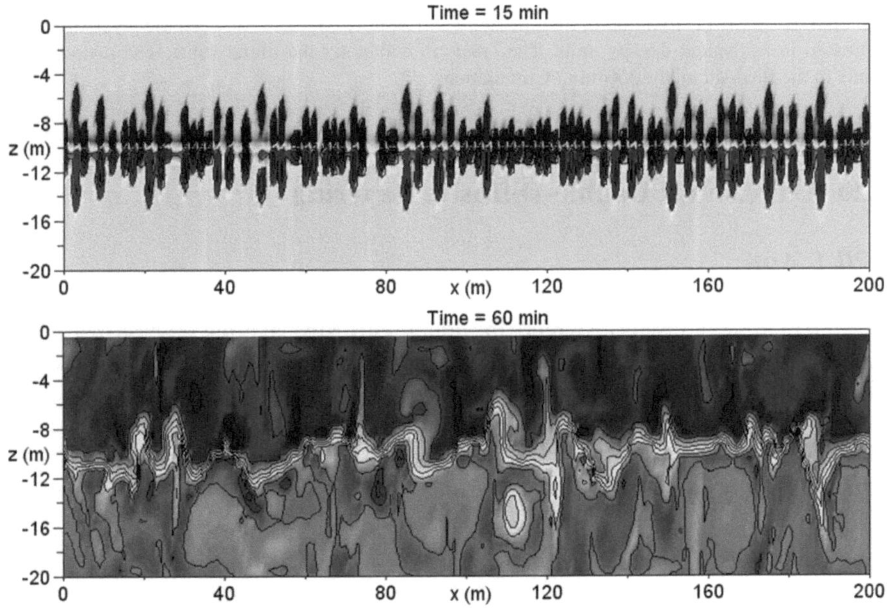

Fig. 3.45 Exercise 12. Snapshots of density distribution (*shading* and *contours*) at selected times of the simulation

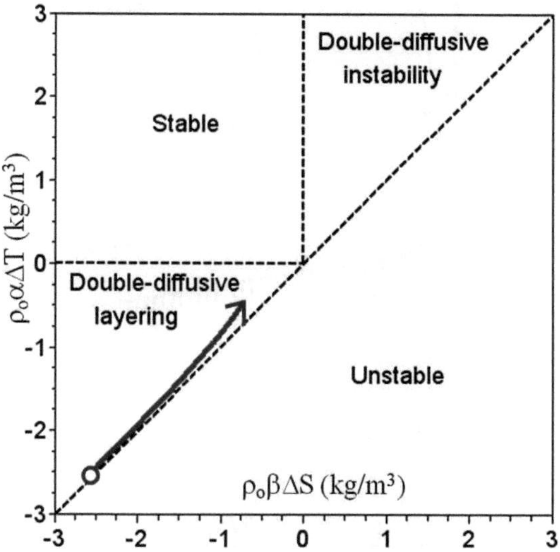

Fig. 3.46 Exercise 12. Evolution of temperature and salinity contrasts between the layers (*arrow*) expressed in equivalent density units. The *open circle* denotes the initial value. The *arrow head* points to the final value after 24 hrs of simulation

3.20.4 Additional Exercises for the Reader

Repeat Exercises 11 and 12 with a two-layer configuration corresponding to a Turner angle in a range of (plus or minus) 60°–75°. Explore how the Turner angle evolves over time. The reader is also encouraged to consider a fluid of three superimposed layers that are density compensated but display contrasts of temperature and salinity.

3.21 Tilted Coordinate Systems

3.21.1 The Governing Equations

The Cartesian coordinate system can be tilted by a certain angle γ (Fig. 3.47) for studies of the dynamics of stratified flows on a sea floor of constant bottom inclination. Since this rotation rotates the entire model domain, inclusion of a free sea surface is not possible anymore, and the rigid-lid approximation (see Sect. 3.7) must be employed. With this approach, the rotated vertical axis z_r differs from the local vertical and, hence, from the orientation of the gravity force. Accordingly, some fraction of the reduced-gravity force appears along rotated x_r surfaces and the

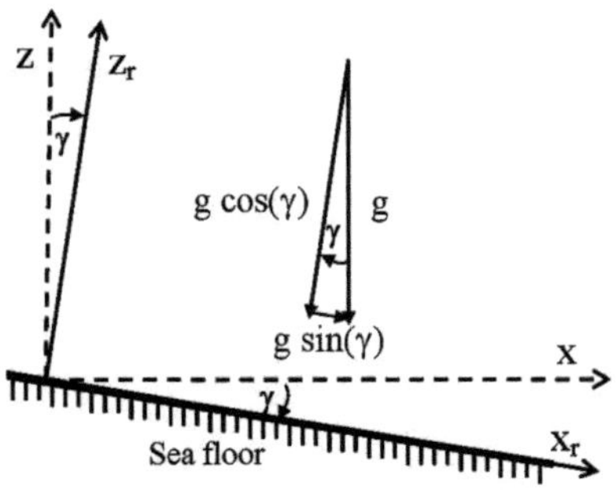

Fig. 3.47 Tilted coordinate system

dynamical equations can be written as:

$$\frac{\partial u_r}{\partial t} + u_r\frac{\partial u_r}{\partial x_r} + w_r\frac{\partial u_r}{\partial z_r} = -\frac{1}{\rho_o}\frac{\partial P}{\partial x_r} + \sin(\gamma)\frac{\rho'}{\rho_o}g + \text{Diff}(u_r) \tag{3.78}$$

$$\frac{\partial w_r}{\partial t} + u_r\frac{\partial w_r}{\partial x_r} + w_r\frac{\partial w_r}{\partial z_r} = -\frac{1}{\rho_o}\frac{\partial P}{\partial z_r} - \cos(\gamma)\frac{\rho'}{\rho_o}g + \text{Diff}(w_r) \tag{3.79}$$

$$\frac{\partial u_r}{\partial x_r} + \frac{\partial w_r}{\partial z_r} = 0 \tag{3.80}$$

$$\frac{\partial \rho}{\partial t} + u_r\frac{\partial \rho}{\partial x_r} + w_r\frac{\partial \rho}{\partial z_r} = \frac{\partial}{\partial x_r}\left(K_h\frac{\partial \rho}{\partial x_r}\right) + \frac{\partial}{\partial x_r}\left(K_z\frac{\partial \rho}{\partial z_r}\right) \tag{3.81}$$

where u_r is the bottom-parallel component of velocity, and w_r is the velocity component perpendicular to the sea floor. The diffusion operator can be defined by:

$$\text{Diff}(\psi) = \frac{\partial}{\partial x_r}\left(A_h\frac{\partial \psi}{\partial x_r}\right) + \frac{\partial}{\partial z_r}\left(A_z\frac{\partial \psi}{\partial z_r}\right)$$

which is only justifiable for small bottom inclinations. As the reader can see, all this coordinate transformation does is to rotate direction-dependent forces such as the reduced-gravity force. The component of the reduced-gravity force normal to the sea floor is sometimes referred to as *buoyant-slope effect*. In case the buoyant-slope effect is fully balanced by a linear bottom-friction force, we yield:

$$\sin(\gamma)\frac{\rho'}{\rho_o}g - ru_r = 0$$

where r is a linear bottom-drag coefficient. This gives an equilibrium flow speed of $u_r = \sin(\gamma)g'/r$, where g' is reduced gravity, which depends on the density excess carried by the flow, bottom inclination and frictional effects. This relation, however, is only useful in situations in which the plume is void of entrainment of ambient fluid. Laboratory experiments suggest that, owing to entrainment, the velocity of the plume head is almost independent of the slope angle (Britter and Linden, 1980). Influence of the Coriolis force, not considered here, can also substantially modify the dynamics. This will take place on time scales exceeding the inertial period.

The above equations are very similar to those used in previous exercises, apart from an additional term in the bottom-parallel component of the momentum equation and a slight modification of the hydrostatic part of pressure when adjusted parallel to the z_r direction. Hence, only minor code modifications are required for simulations of density-driven flows on a sloping sea floor.

3.22 Exercise 13: Stratified Flows on a Slope

3.22.1 Aim

The aim of this exercise is to employ a tilted Cartesian coordinate system to simulate the dynamics and instability of a stratified flow on a sea floor of uniform slope.

3.22.2 Task Description

Consider a tilted model domain, 500 m in length and 100 m in thickness, resolved by grid spacings of $\Delta x = 5$ m and $\Delta z = 2$ m (Fig. 3.48). The sea floor has a bottom inclination of 5°. Lateral boundaries are cyclic. The ocean is initially at rest.

The simulation is started with prescription of a 20-m thick near-bottom layer of a density of 1,028.2 km/m^3. The ambient ocean has a density of 1,028.0 km/m^3. Small random noise is added to this density field. Eddy diffusivities and viscosities are set to a uniform value of 10^{-2} m^2/s. The bottom-drag coefficient is set to $r = 0.001$. The total simulation time is 2 hrs with data outputs at 1-min intervals. The time step is set to $\Delta t = 1$ s. The pressure accuracy for the S.O.R. iteration is set to $\epsilon = 0.001$ Pa.

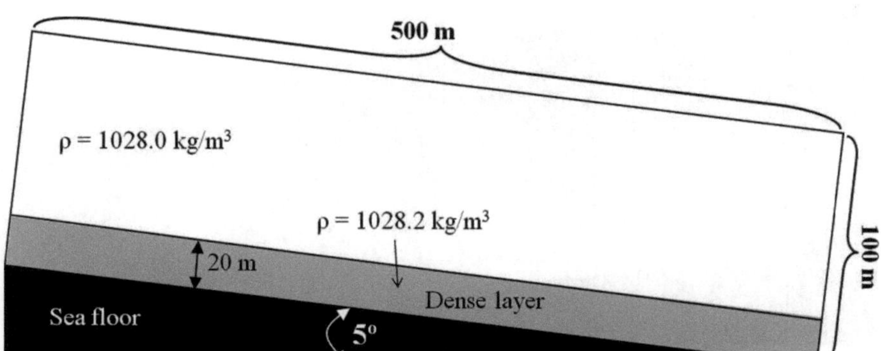

Fig. 3.48 Exercise 13: Initial configuration of the density field

3.22.3 Results

As expected, the dense bottom layer gradually accelerates on the sloping bottom (Fig. 3.49). This creates a vertical shear flow. The shear flow becomes subject to the Kelvin-Helmholtz instability process (see Sect. 3.10). Dynamical disturbances appear after 70 min of simulation and create counter-clockwise rotating vortices. At this stage, the near-bottom flow has reached speeds of 80 cm/s.

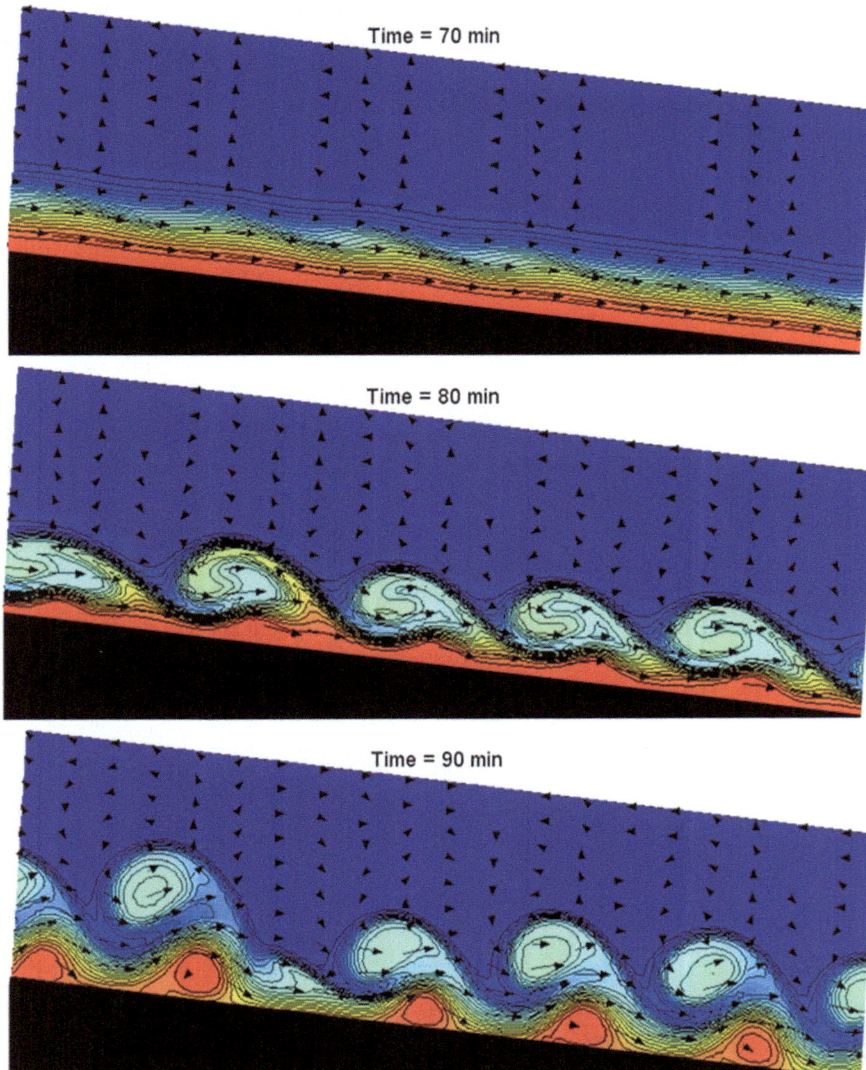

Fig. 3.49 Exercise 13: Density distributions (*color shading* and *contours*) at selected times of the simulation. *Red shading* refers to the densest water

The dynamic instability produces vigorous vertical mixing via the breaking of internal waves. Interestingly, at later times of the simulation, velocity shear owing to bottom friction creates a series of clockwise-rotating vortices that "roll" downward on the slope and interact with previously created vortices that have detached from the sea floor.

3.22.4 Additional Exercise for the Reader

Conduct a reference experiment with a small bottom inclination of, say, 1 degree. Explore whether this slope is sufficient to create dynamical instabilities. If not, increase the bottom inclination in additional experiments until the vertical shear of the near-bottom flow is at a level to produce dynamical instabilities.

3.23 Estuaries

3.23.1 Definition

Estuaries are semi-enclosed and narrow bodies of water that are connected to the coastal ocean, at least intermittently, and in which the salinity is measurably different from that in the adjacent ocean. Estuaries are categorised as two different types (Fig. 3.50).

Positive estuaries have a salinity less that of the adjacent ocean. These estuaries are exposed to a humid climate and experience a surplus of freshwater input via rivers over evaporation. Positive estuaries can be characterised by a two-layer structure with light, low-salinity water flowing toward the ocean in the upper layer and inflow of seawater in the layer underneath. The region of maximum lateral salinity gradients is called *brackish water.*

Negative estuaries, also called inverse estuaries, are situated in arid climatic zones and they experience a net water loss owing to an excess of evaporation over

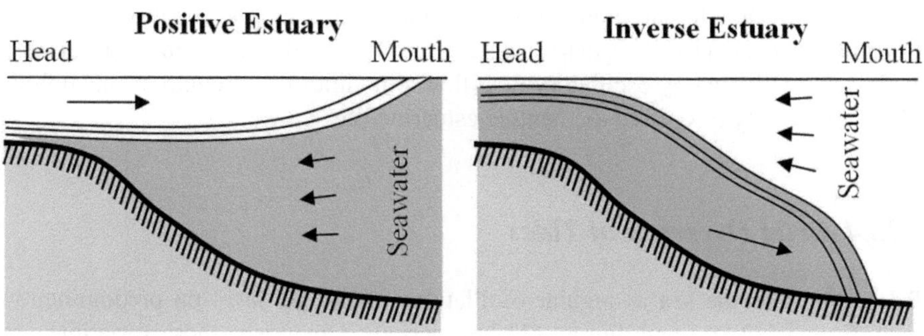

Fig. 3.50 Schematic of the circulation in positive and inverse estuaries

precipitation-related inflows. Inverse estuaries are characterised by an outflow of hypersaline bottom water and inflow of ambient seawater in the surface layer.

3.23.2 Classification of Estuaries According to Origin

According to their geological characteristics, estuaries can be classified as:

- drowned river valleys or coastal-plain estuaries
- bar-built estuaries or lagoons
- fjords
- tectonically caused estuaries

Coastal-plain estuaries and drowned river valleys were formed by the gradual rise of sea level after the last glacial period, 10,000 years ago. Lagoons are similar to coastal-plain estuaries, but these are usually situated parallel to the coastline. Many lagoons have narrow outlets and minimum freshwater inflow, often creating hypersaline conditions.

Fjords are long, narrow inlets with steep sides, created in a valley carved by glacial activity. Most fjords are deeper than the adjacent sea with depths of 1,000–2,000 m. Fjords generally have a sill at their mouth which limits the exchange with the ambient sea, but in many cases creates extreme tidal flows. For instance, Saltstraumen in Norway involves water currents of up to 10 m/s in speed and is therefore often described as the world's strongest tidal current.

Tectonic estuaries include those created by landslides, faulting and volcanic eruptions. San Francisco Bay in the USA is an example.

3.23.3 The Dynamics of Positive Estuaries

Continental runoff is the principle driver of flows in positive estuaries. This runoff is associated with a barotropic pressure gradient directed toward the sea aiming to push low-salinity surface water seaward. The density interface between the layers, however, gives rise to baroclinic pressure gradients that support a bottom inflow of ambient seawater into the estuary (see Fig. 3.50). The degree of mixing between the layers, facilitated by oscillatory tidal flows, modifies the strength of this inflow, which can supply oxygen to the benthic estuarine life.

3.23.4 Brief Overview of Tides

Tides appear in the sea as regular oscillations of the sea level on predominantly semi-diurnal (12 hrs) and diurnal (24 hrs) periods. Obviously, these variations are caused by convergence and divergence of oscillatory lateral flow. *High tide* is the instance of maximum sea level. *Low tide* refers to the minimum sea level. The *tidal range* is the difference between high tide and low tide. *Slack water* refers to times of no or only little tidal flow.

The tide-generating force involves the gravitational attractive force between the Earth/Moon and Earth/Sun systems and the centrifugal force. The magnitude of the gravitational force is proportional to mass and decreases rapidly with the square of distance between bodies. Therefore, the Moon (due to close distance) and the Sun (due to large mass) are the only contributors to tides on Earth. Tides are created owing to a slight imbalance of these forces at the sea surface.

3.23.5 Dynamic Theory of Tides

Oceanic tides are long surface gravity waves of a phase propagation being constrained by the well-known dispersion relation Eq. (3.8). Owing to this constraint and the existence of coastlines, tidal oceanic waves simply cannot keep pace with the progress of the astronomical forcing. Instead of this, the tidal signal in the ocean is rather of the form of a standing-wave response that varies in magnitude according to the regional geometry of an oceanic basin. This "dynamic theory of tides", first proposed by Pierre-Simon Laplace in 1775 (Laplace, 1775), successfully describes the apparently complex tidal patterns in the ocean.

The influence of the Coriolis force on tidal motion gives rise to a Kevin-wave response (see Sect. 3.6.2 of Kämpf (2009)). Rotational effects support maximum tidal sea levels propagating along the coast. *Amphidromic points* are locations that display no tidal range a tidal constituent (see below). Radiating from such a point are *co-tidal phase lines*, which connect points where high tide occurs simultaneously. *Co-tidal range lines*, on the other hand, connect points of the same tidal range. To this end, the tidal signal in the ocean propagates around amphidromic points and maximum tidal ranges are often found along coastlines. This makes coastal regions vulnerable to storm surges if these coincide with a high tide.

3.23.6 Tides in Estuaries

Tidal co-oscillations in the estuary follow from the tidal forcing of the adjacent ocean. As the tidal wave approaches shallower water, its wavelength decreases according to Eq. (3.8). Dependant on geometry and because of the shortening of the tidal wave, some estuaries exhibit enormous tidal sea-level variations. The most amplified response occurs if a wave node establishes close to the estuary's entrance. The Bay of Fundy on the east coast of North America, for instance, supports a tidal range of 17 m at the northern end of the Minas Basin, which makes boating and fishing activities rather transient ones. Natural oscillations in an enclosed or partially enclosed body of water are called *seiches*.

3.23.7 Tidal Patterns

Oceanic tides are the composite of a set of several *tidal constituents* also called *partial tides*. The most prominent tidal constituents are briefly described in the following. The Moon creates a semidiurnal lunar tide (abbreviated as M2), with a period of 12 hrs and 25 min, and a diurnal lunar tide (O1), with a period of 24 hrs and 50 min. The Sun causes a semidiurnal solar tide (S2), with a 12-hrs period, and a diurnal solar tide (K1), with a 24-hrs period. The M2 constituent is often dominant.

The longer period of semi-diurnal and diurnal lunar tides compared with solar tides is caused by progression of the Moon's location during one full rotation of the Earth. Dependant on geometry, tides can appear in shelf seas and estuaries as either *semidiurnal tides*, *diurnal tides*, or a mixture of the latter two, called *mixed tides*. Tides in the Bay of Fundy, for instance, are mixed tides, but diurnal sea-level variations appear more dominant than semi-diurnal variations. The *form factor* compares the amplitudes of semi-diurnal tidal constituents (M2 and S2) with those of diurnal tides (O1 and K1). These amplitudes can be derived from field observations using a method called *harmonic analysis*, not replicated here. The form factor tells the tidal characteristics of a region.

Wave interference is the partial or full offset of the sea-level elevation of one wave pattern by a sea-level depression of another wave pattern. Interference of solar and lunar tides gives rise to regular variation of the tidal range between *spring tide*, when it has its maximum, and *neap tide*, when it has its minimum. The period between spring and neap tide is about 14 days. In certain regions, such as Gulf St. Vincent and Spencer Gulf in South Australia, this interference is almost perfect and leads to almost absent tidal flows during the neap tide. This phenomenon is called *dodge tide*, first reported by the explorer Matthew Flinders in 1814 (Flinders,1814).

3.23.8 Classification of Estuaries According to Stratification and Circulation

The circulation in positive estuaries depends on the amount of freshwater received from continental runoff and the magnitude of tidal flows inducing vertical mixing. When using the period of the predominant tidal constituent for reference, estuaries can be classified in terms of the amount of freshwater received over this period R, called freshwater volume, and the amount of water that the tidal flow moves into an estuary over a tidal cycle, referred to as V in the following. The tidal volume can be calculated from tidal range observations and the surface area of an estuary and it is a measure of the magnitude of tidal flows in a region. The freshwater volume has to be derived from measurements of flow rates at the mouth of individual rivers, also called sub-estuaries, feeding into the parent estuary.

Salt-wedge estuaries are classified by an R/V ratio exceeding unity. The surface water has small salinities along the entire length of the estuary and this becomes largely undiluted exported into the adjacent sea. A wedge of salty oceanic water is found underneath the low-salinity water water. Except for the sharp density interface between the surface and bottom layers, there are little salinity differences in each of the layers. Examples of salt-wedge estuaries include the world's mightiest rivers, the Conga, Amazon, and Mississippi Rivers, but this type of estuary can also develop in much smaller rivers.

Highly stratified estuaries are classified by a slightly smaller R/V ratio of values >0.1. Here, substantial amounts of saltier water are entrained into the surface layer outflow, such that both salinity and flow rate of this outflow increase along the length of the estuary. The bottom layer consists of largely undiluted seawater, but the loss of saline water to the surface layer implies a continuous bottom inflow of seawater from the adjacent sea. The Hardanger Fjord in Norway, for example, turns seasonally into a highly stratified estuary whereby the freshwater source is provided by summertime melt of glaciers. Constrained by the existence of a sill, however, the inflow of seawater is often not dense enough to replenish bottom waters of a fjord. As a consequence of this, bottom waters can become depleted in dissolved oxygen (anoxic) as dead organic matter decays in the bottom layer. In some fjords, this situation can last for several years.

A further decrease of the R/V ratio to values >0.01 leads to establishment of *slightly stratified estuaries*. The relative influence of tidal mixing is enhanced and, in contrast to entrainment being a one-way process, this creates mixing between both layers. Hence, the salinity increases in both layers along the length of the estuary with a top-to-bottom salinity difference remaining approximately constant. Examples of slightly stratified estuaries include Chesapeake Bay situated in the north-western Atlantic Ocean. Maximum river flow in the spring supports a stronger density interface preventing the fresh surface water and saltier bottom water from mixing. Owing to severe shortage of dissolved oxygen, major kills of commercially important bottom-dwelling animals occur during this time.

Positive R/V values <0.001 characterise a situation in which tidal mixing is strong enough to suppress density stratification most of the time. This regime characetrises *vertically mixed estuaries*. Salinity at any point of the estuary is almost uniform from surface to bottom and salinity increases from head to mouth of the estuary. The Bay of Fundy is an example of a vertically mixed estuary.

Finally, a negative R/V ratio corresponds to a net water loss of an estuary which is characteristic of *inverse estuaries*. Examples of inverse estuaries are Spencer Gulf and Gulf St. Vincent in South Australia.

3.23.9 Transport Timescales in Estuaries

Owing to high human population density, most estuaries are subject to various kinds of pollution such as discharges of sewage and other wastewater, or oil spills. Hydrodynamic models are frequently applied to study the dilution and dispersal of pollutants in estuaries. Various timescales have been introduced by scientists for this purpose. These timescales are briefly described in the following.

Residence time is the time it takes for a virtual water parcel to escape from a given area. Particle-tracking methods, using Lagrangian floats such as in Exercise 9, are used to calculate this timescale. For the circulation shown in Fig. 3.51, for instance, the residence time of a water parcel released at point A is the time it takes until it reaches the exit boundary at point B. In this sense, a map of residence times can be constructed for the entire region of interest. With consideration of parcels entering the region from outside, it is also possible to calculate a mean transit time including standard variation for a region of interest. Knowledge of residence times are useful for management of oil spills and identification of "shadow" regions of little flow.

Flushing time is the time it takes for the water volume of a given region to be (almost) fully replaced by ambient water. Flushing times are computed using Eulerian tracer fields such as in Exercise 9. To this end, tracer concentrations of unity are initially allocated to the region of interest whereas concentrations are kept at zero value outside during the simulation. Flushing times can then be estimated as the time is takes until concentration has dropped below a certain threshold value usually taken at $\exp(-\pi) \approx 0.04$. This implies that a region is considered flushed when about 96% of its initial water has been replenished with waters from a pre-defined source region. The resultant flushing time distribution depends on the start time of the simulation. Distributions of flushing times are useful to illustrate regions that are relatively stagnant. For the situation displayed in Fig. 3.51, for example, we can anticipate delayed flushing along the coastal zones and inside the centre of the eddy in Fig. 3.51.

Age of a virtual water parcel or water volume is the time elapsed since it has entered the system (Deleersnijder et al., 2001). When formulated by means of Lagrangian floats, age tracking is similar to the calculation of residence time. When using a large number of floats, which can be computationally "expensive", the Lagrangian method reveals age distributions within a grid cell as a function of both location and time. Nevertheless, water age is usually calculated from Eulerian concentration fields according to the modified advection-diffusion equation:

$$\frac{\partial A}{\partial t} + \text{Adv}(A) = \text{Diff}(A) + 1 \tag{3.82}$$

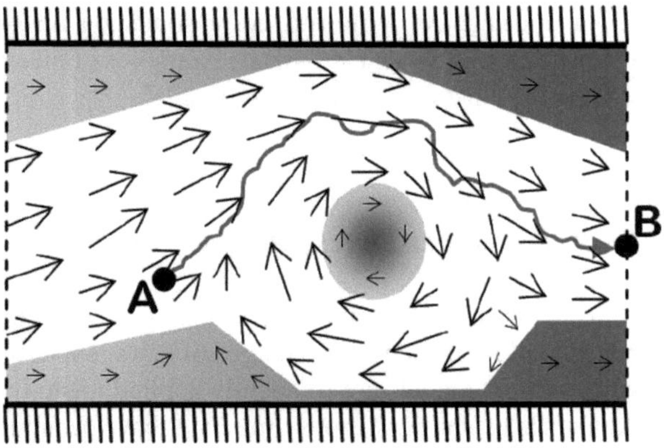

Fig. 3.51 Schematic for explanation of residence time and flushing time. *Shaded areas* represent regions of delayed flushing

where A is grid-averaged age, Adv(A) and Diff(A) denote the advection and diffusion terms, and the constant unity represents ageing. In a semi-enclosed body of water such as an estuary, entrainment of younger ambient seawater leads to establishment of an equilibrium age distribution (see Fig. 3.52). As flushing times, the spatial distribution of age indicates the *marine connectivity* in a region of interest. The advantage of predicting age rather than flushing time is that it does not require any threshold value and that it allows for continuous prediction without the need to restart the model in different seasons of the year.

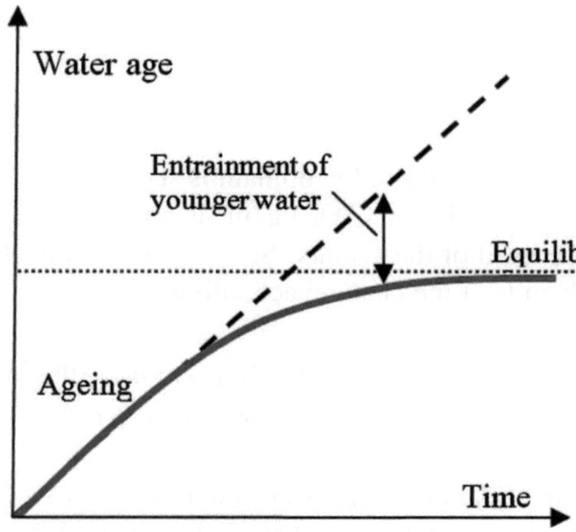

Fig. 3.52 Entrainment of younger water leads to establishment of a steady value of age

3.24 Exercise 14: Positive Estuaries

3.24.1 Aim

The aim of this exercise is to simulate the circulation and density structure of a positive estuary subject to varied magnitudes of freshwater inflow and exposed to tidal flows of a given tidal range. To make the simulation more realistic, variable channel width is implemented in the vertical ocean-slice model.

3.24.2 Task Description

We consider a river channel of 120 km in length and 50 m in width opening up to the sea (Fig. 3.53). The channel's sidewalls are for simplicity vertical. The channel widens to 500 m at its opening to the sea. Channel depth increases linearly from 5 to 20 m from head to mouth. We can employ the 2d vertical ocean slice model with a few modifications (see below) under the assumption that the flow is uniform across the channel width. A vertical grid spacing of 1 m together with a horizontal grid spacing of 2 km are used.

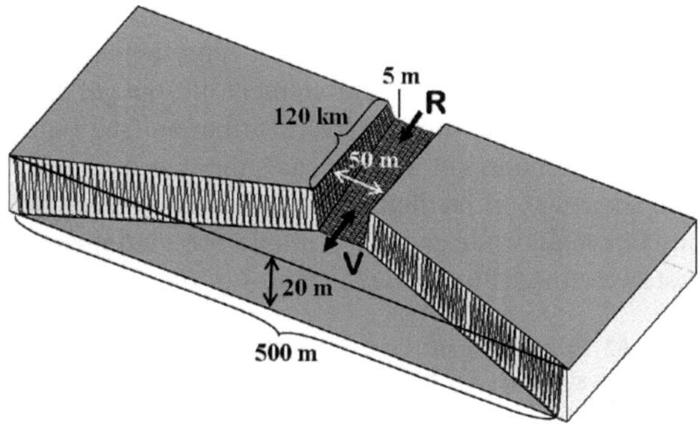

Fig. 3.53 Exercise 14. Model configuration. The freshwater inflow is characterised by a certain freshwater volume R. Tides are related to a certain tidal volume V. See text for explanation

Initially, density in the domain is set to an oceanic value of $1{,}027\,\text{kg/m}^3$. Freshwater of a density of $1{,}000\,\text{kg/m}^3$ is prescribed throughout the entire water column near the head of the estuary. Sea level variations, forcing the model, are prescribed at both ends of the channel according to:

$$\eta_{\text{head}} = \delta\,[\eta_{\text{river}} - \eta_{\text{tide}} \sin\{2\pi(t + t_o)/T\}]$$
$$\eta_{\text{mouth}} = -\eta_{\text{tide}} \sin(2\pi t/T)$$

where the tidal amplitude (half the tidal range) is kept at a value of $\eta_{\text{tide}} = 0.25\,\text{m}$, tidal period T is taken as 12 hrs, t_o is a phase difference of tidal amplitudes between head and mouth, taken as 30 min, and η_{river} denotes the elevation of the river above sea level that drives freshwater into the estuary. The latter is varied between 0.01 and 0.5 m. The δ parameter is gradually adjusted from zero to unity over the initial 6 hrs of simulation. This is required to avoid the creation of unwanted initial disturbances of the form of long surface gravity waves. The resultant tidal flows create a tidal range of 0.5–1 m along the channel. Zero-gradient conditions are used for all other variables at open boundaries. As in Exercise 3, the sea-level variations imposed are converted to dynamic pressure variations.

An advanced diagnostic turbulence scheme, first proposed by Pacanowski and Philander (1981), is adopted for calculation of eddy viscosity and eddy diffusivity. Details are given below. Horizontal eddy viscosity/diffusivity is set to a uniform value of $0.1\,\text{m}^2/\text{s}$ and a bottom friction parameter of $r = 1 \times 10^{-3}$ is used.

Additionally, water age is calculated using Eq. (3.83) to identify regions sensitive to oxygen depletion. Initially, water age is set to zero in the entire model domain. There are two main oxygen sources in an estuary: fluxes across the sea surface and inflow of well-oxygenated water from the ambient sea. To capture these sources, age is kept at zero values in all surface grid cells and outside the mouth of the estuary.

The total simulation time of experiments is 20 days. In addition to snapshot outputs on a two-hourly basis, the author decided to include outputs of data averaged over each tidal cycle. The numerical time step is set to 20 secs. A pressure accuracy of $\epsilon = 10^{-3}$ Pa is used for the S.O.R. scheme.

3.24.3 Implementation of Variable Channel Width

The width b of the channel varies only in the x-direction and not with depth. Wu (2007) presents dynamical equations for more complex river shapes. Horizontal and vertical velocities are width-averaged across the channel. Under these assumptions, the momentum equations and advection-diffusion equation for scalars remain the same except for a modification of the diffusion term. This term is now given by:

$$\text{Diff}(\psi) = \frac{1}{b}\frac{\partial}{\partial x}\left(b\,K_h \frac{\partial \psi}{\partial x}\right) + \frac{\partial}{\partial z}\left(K_z \frac{\partial \psi}{\partial z}\right) \tag{3.83}$$

where ψ is a substitute for variables, $b(x)$ is channel width, and K represents either eddy viscosity in the momentum equations or eddy diffusivity for scalars. The continuity equation for the width-averaged flow can be written as:

$$\frac{\partial(b\,u)}{\partial x} + \frac{\partial(b\,w)}{\partial z} = 0 \tag{3.84}$$

and vertical integration leads to:

$$\frac{\partial \eta}{\partial t} = -\frac{1}{b}\frac{\partial(b\,h\,\langle u\rangle)}{\partial x} \tag{3.85}$$

where $\langle u\rangle$ is horizontal flow velocity averaged over both depth and width of the channel. Owing to the appearance of channel width in the continuity equation, coefficients in the S.O.R. scheme (see Sect. 3.4) are now given by:

$$a_e = b_e\,\Delta z/\Delta x \quad , \quad a_w = b_w\,\Delta z/\Delta x \quad , \quad a_t = b_k\,\Delta x/\Delta z \quad , \quad a_b = b_k\,\Delta x/\Delta z$$

where

$$b_e = 0.5\,(b_k + b_{k+1}) \quad \text{and} \quad b_w = 0.5\,(b_k + b_{k-1})$$

Accordingly, the source term on the right-hand side of (3.24) is given by:

$$q^*_{i,k} = \frac{\rho_o}{\Delta t}\left[\left(b_e\,u^*_{i,k} - b_w\,u^*_{i,k-1}\right)\Delta z + b_k\left(w^*_{i,k} - w^*_{i+1,k}\right)\Delta x\right] \tag{3.86}$$

3.24.4 Advanced Turbulence Closure

The vertical mixing scheme by Pacanowski and Philander (1981) is a sole function of the Richardson number (Eq. 3.61). This scheme has been developed for tropical-ocean applications, but we take the freedom to adopt this scheme for this exercise. Vertical eddy viscosity is calculated from

$$A_z = \frac{\nu_1}{(1 + 5\ Ri)^2} + \nu_o \tag{3.87}$$

with $\nu_1 = 5 \times 10^{-3}\ \mathrm{m^2/s}$ and $\nu_o = 5 \times 10^{-4}\ \mathrm{m^2/s}$. Eddy diffusivity is parameterised as:

$$K_z = \frac{A_z}{(1 + 5\ Ri)} + k_o \tag{3.88}$$

where k_o is set to $1 \times 10^{-5}\ \mathrm{m^2/s}$. For the calculation of the Richardson number it is useful to recall that, with use of the Arakawa C-grid, eddy viscosity/diffusivity values are determined at scalar grid points which are centred between velocity grid points. To this end, the Richardson number is derived from:

$$Ri^n_{i,k} = -2\Delta z \frac{g}{\rho_o} \frac{(\rho^n_{i-1,k} - \rho^n_{i+1,k})}{(u_t - u_b)^2} \tag{3.89}$$

where

$$u_t = 0.5(u^n_{i-1,k} + u^n_{i-1,k-1}) \text{ and } u_b = 0.5(u^n_{i+1,k} + u^n_{i+1,k-1})$$

There is no need to account for convective mixing in this exercise.

3.24.5 Results

The tidal volume flux is about $1.4\ \mathrm{m^3/s}$ per unit width of the estuary for all forcing cases. The choice of $\eta_{\text{river}} = 1$ cm leads to an R/V ratio of 0.01 and therefore corresponds to a vertically-mixed estuary (Fig. 3.54, top panel). The brackish water zone does not reach far into the estuary and the transition zone between freshwater and seawater occupies only a width of 10–15 km. The lack of density stratification provides a means of oxygen supply (also called *ventilation*) via surface fluxes over the entire length of the estuary.

In contrast, an increased value of $\eta_{\text{river}} = 30$ cm, corresponding to an R/V ratio of 1.9, creates a salt-wedge estuary characterised by a freshwater surface layer extending the entire length of the estuary and spilling into the ambient sea (Fig. 3.54, bottom panel). In this situation, the brackish water zone is located closer to the estuary mouth, which enables relatively rapid ventilation of the bottom layer through the baroclinic return flow despite the existence of strong density stratification.

Forcing with $\eta_{\text{river}} = 5$ cm gives an R/V ratio of 0.16 and leads to the establishment of a highly stratified estuary (Fig. 3.55, top panel). Saltier bottom water is entrained into the surface outflow. Owing to turbulence reduction via density stratification, bottom layers attain an age of 13–14 days (Fig. 3.55, bottom panel), which is an indication that oxygen depletion might occur in this zone. Discharge of pollutants should be avoided in such regions of little ventilation.

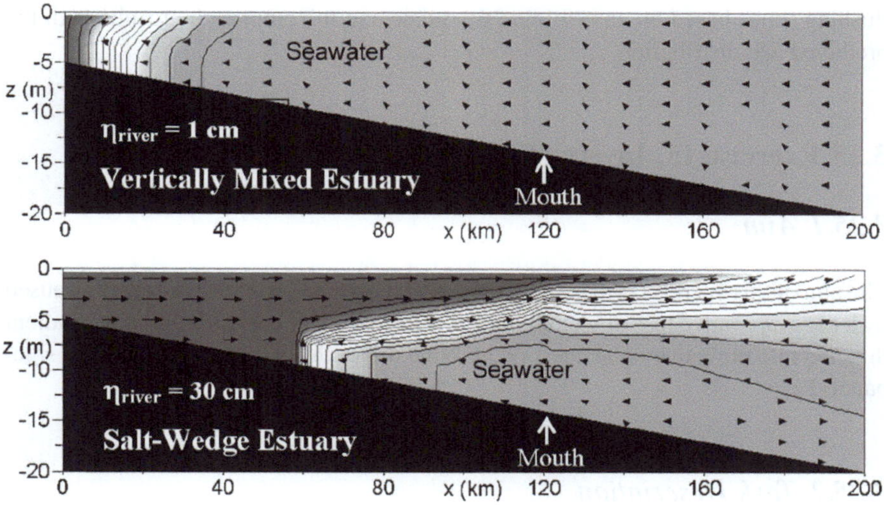

Fig. 3.54 Exercise 14. Tidally averaged density distribution (*shading* and *contours*) after 20 days of simulation for different forcing cases. Arrows display averaged flow vectors

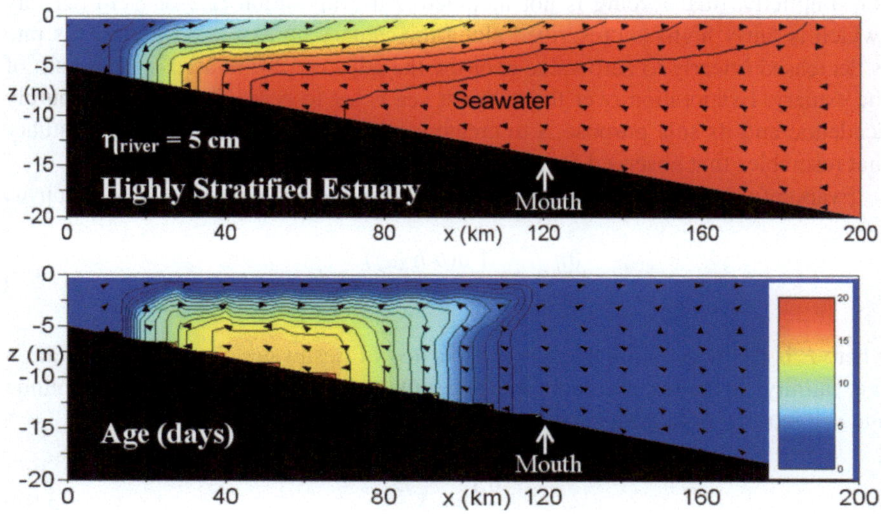

Fig. 3.55 Exercise 14. Results for a highly stratified estuary. *Top panel*: Tidally averaged density distribution (*color shading* and *contours*). *Blue* (*red*) *shading* denotes freshwater (seawater). *Bottom panel*: Distribution of water age (days). Arrows display averaged flow vectors

3.24.6 Additional Exercises for the Reader

The reader is encouraged to add some bathymetric variations to the sea floor (such as a sill near the estuary's mouth) and to explore changes in both the dynamics and water age distributions. The reader should also try to calculate flushing times based on an e-folding timescale and to compare the results with the predicted age distribution.

3.25 Exercise 15: Inverse Estuaries

3.25.1 *Aim*

The aim of this exercise is to explore the circulation of inverse estuaries being caused by a net evaporative loss of water. Since this loss of water creates a pressure gradient directed into the estuary, the question is: What drives the outflow (see Fig. 3.50, right panel)?

3.25.2 *Task Description*

The model configuration is identical to that of the previous exercise with a few modifications outlined in the following. The estuary is closed on the left side and, for simplicity, tidal forcing is not included. An evaporation rate of 5 cm per day (which is fairly high) is prescribed along the estuary up to x = 110 km. This rate is decreased linearly to zero value over the the adjacent 20 km across the mouth of the estuary. Evaporation over the ambient sea is set to zero. This design is purely academic and its sole purpose is to create a salinity distribution along the estuary that resembles that observed in real inverse estuaries.

Evaporative water loss appears in the vertically integrated continuity equation as:

$$\frac{\partial \eta}{\partial t} = -\frac{1}{b}\frac{\partial (b\ h\ \langle u \rangle)}{\partial x} - E \tag{3.90}$$

where E is the evaporation rate. Owing to a loss of freshwater, the salt concentration (i.e. salinity) in the water column increases. On the basis of conservation of volume and salt mass, this salinity change can be calculated from:

$$\frac{\partial S}{\partial t} = S\frac{E}{h} \tag{3.91}$$

where h is the depth over which this salinity increase becomes distributed. We use a linearised equation of state expressed in terms of density anomaly ρ':

$$\rho' = \rho_o \beta (S - S_o) \tag{3.92}$$

where temperature effects are ignored, the salinity coefficient β is taken as 8×10^{-4}, and S_o is a reference salinity, taken as 35 g/kg. Accordingly, Eq. (3.91) can be converted into a density change according to:

$$\frac{\partial \rho'}{\partial t} = \left(\rho_o \beta S_o + \rho'\right) \frac{E}{h} \tag{3.93}$$

In the model, this density change is distributed in the uppermost grid cell ($h = \Delta z$). Convection by unstable density stratification can mix this salinity anomaly to deeper layers. Owing to coarse grid spacing, convective mixing needs to be parameterised. This is done here via an increase of eddy diffusivity to a value of $k_z = 0.01\,\text{m}^2/\text{s}$. The total simulation time is 20 days. The evaporative forcing is applied only for the first 10 days and thereafter disabled. Reasons for this treatment will become obvious with inspection of the results.

3.25.3 Results

Due to evaporation, salinity increases toward the head of the estuary, as anticipated (Fig. 3.56, top panel). Maximum density anomalies are $3\,\text{kg/m}^3$, which converts to a salinity anomaly of 3.65 g/kg, using Eq. (3.92). Surprisingly, an overturning circulation as sketched in Fig. 3.50 (right panel) does not establish during times of evaporation. Instead, there is only a weak depth-independent (barotropic) flow running into the estuary and replacing the water volume lost through evaporation.

The lack of baroclinic flow is contrary to the anticipation that the salinity increase and, hence, density increase in an inverse estuary directly drives a hypersaline bottom outflow. This does not happen during the first 10 days of simulation here because the evaporative water loss in the estuary creates a barotropic pressure gradient directed into the estuary which operates to override any density effects. In this situation, the barotropic inflow manages to balance both the water loss and the salinity increase associated with evaporation.

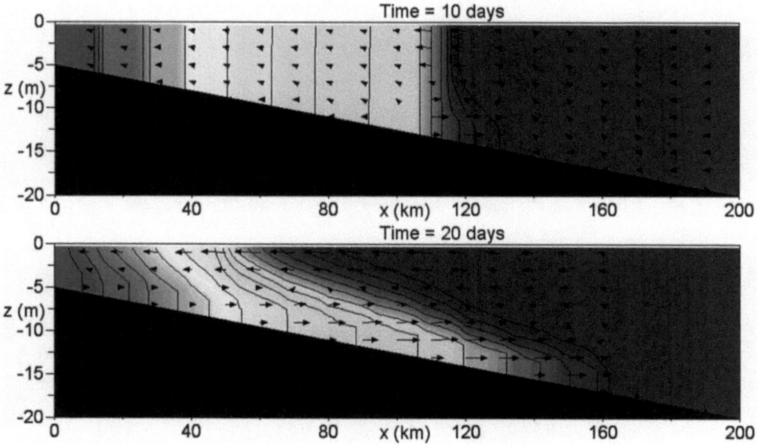

Fig. 3.56 Exercise 15. Density distributions (*shading* and *contours*) after 10 and 20 days of simulation. Arrows display averaged flow vectors

On the other hand, with disappearance of the evaporative forcing, inertia leads to rapid levelling out of the sea level while lateral density gradients remain. Baroclinic pressure gradients are left behind and now can create the outflow of hypersaline bottom water into the ambient sea (Fig. 3.56, bottom panel). The sea level in the estuary drops owing to this outflow and triggers an intensified inflow in the surface layer. Findings of this exercise indicate that the circulation in inverse estuaries is intermittent and responds rapidly to variations in evaporation rates.

3.25.4 Additional Exercise for the Reader

Describe a temporally varying evaporation rate according to:

$$E(t) = E_o\,[1 - \cos(2\pi t/T)]$$

where $E_o = 5\,\text{cm/day}$. Vary the period T between 1 and 10 days and explore the resultant exchange circulations.

Chapter 4
2.5D Vertical Slice Modelling

Abstract This chapter introduces the reader to 2.5-dimensional modelling in a vertical ocean slice which allows for inclusion of the Coriolis force. Exercises address geostrophic adjustment of density fronts, coastal upwelling and Ekman pumping. The last exercise of this chapter explains the curiosity that winds can create flows in the ocean running opposite to the wind direction.

4.1 The Basis

4.1.1 Adding Another Half Dimension

Elongated dynamical features being influenced by the Coriolis force, such as oceanic fronts, can be described to first-order approximation by the dynamics in a vertical ocean slice with vanishing gradients of all variables normal to this slice (Fig. 4.1). This approach is called the *2.5-dimensional vertical ocean-slice model*.

4.1.2 The Geostrophic Balance

Processes considered in this section involve the Coriolis force. Therefore, we briefly revisit the fundamentals of flows dominated by a balance between the Coriolis force and the horizontal pressure-gradient force; that is, the *geostrophic balance*. The momentum equations for pure geostrophic flow in Cartesian coordinates are given by:

$$-f v_{geo} = -\frac{1}{\rho}\frac{\partial P}{\partial x} \tag{4.1}$$

$$+f u_{geo} = -\frac{1}{\rho}\frac{\partial P}{\partial y} \tag{4.2}$$

where f is the Coriolis parameter (see Sect. 2.1). Geostrophic flows run along lines of constant pressure, called *isobars*. With inclusion of the hydrostatic balance, which

J. Kämpf, *Advanced Ocean Modelling*, DOI 10.1007/978-3-642-10610-1_4,

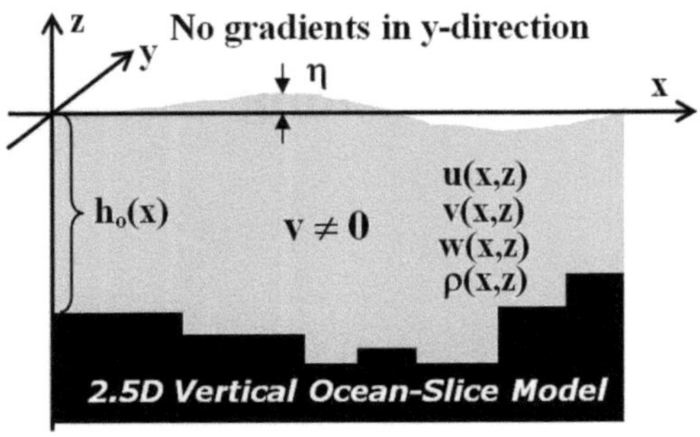

Fig. 4.1 The 2.5d vertical ocean slice

is valid for shallow-water processes (i.e. processes of a horizontal scale exceeding their vertical scale by far), the latter equations can be formulated as:

$$\frac{\partial v_{\text{geo}}}{\partial z} = +\frac{g}{\rho f}\frac{\partial \rho}{\partial x} \tag{4.3}$$

$$\frac{\partial u_{\text{geo}}}{\partial z} = -\frac{g}{\rho f}\frac{\partial \rho}{\partial y} \tag{4.4}$$

These relations, called the *thermal-wind equations*, imply that geostrophic flow displays a vertical shear in the presence of lateral density gradients. In oceanography, the application of the thermal-wind equations for derivation of the baroclinic geostrophic flow field from density measurements is called the *geostrophic method*. See Pond and Pickard (1983) for a detailed description of this method.

Geostrophic surface currents, not captured by the geostrophic method, follow the relationships:

$$-f v_{\text{geo}} = -g\frac{\partial \eta}{\partial x} \tag{4.5}$$

$$+f u_{\text{geo}} = -g\frac{\partial \eta}{\partial y} \tag{4.6}$$

and run along lines of the same sealevel elevation. Sea-level anomalies derived from satellite altimetry can therefore be used to map the surface circulation of the ocean. In the case of uniform density, the latter equations describe depth-independent geostrophic flow.

The *beta-plane approximation* describes departures of the Coriolis parameter from a constant value and is given by:

$$f = f_o + \beta y \tag{4.7}$$

where β is the meridional variation of the Coriolis parameter with a value of $\beta = 2.2 \times 10^{-11}$ m^{-1}s^{-1} at mid-latitudes, and y is the distance in metres with respect to the centre of the Cartesian coordinates system defining f_o. Note that y becomes negative for locations south of this centre.

Horizontal divergence of geostrophic flow is given by:

$$\frac{\partial u_{\text{geo}}}{\partial x} + \frac{\partial v_{\text{geo}}}{\partial y} = -\frac{\beta}{f} v_{\text{geo}} \tag{4.8}$$

On small spatial scales <100 km, this divergence is negligibly small. As a consequence of this, barotropic geostrophic flow tends to follow bathymetric contours – a feature referred to as *topographic steering*. On larger scales of ocean basins (~1,000 km), meridional (north–south) flow creates a small divergence that plays an important role in the wind-driven circulation of the oceans balancing the wind-induced flow divergence in the surface Ekman layer. This balance is known as the *Sverdrup balance*.

4.1.3 Scaling

The temporal Rossby number, defined by Eq. (2.7), compares the inertial period with the time scale of a time-variable process. Coriolis effects become important for a temporal Rossby number of values less than unity or, in other words, if the time scale of a process exceeds the inertial period. Stationary processes, on the other hand, are controlled by the Coriolis force if the Rossby number (without prefix) is small compared with unity. The Rossby number is defined by:

$$Ro = \frac{U}{fL} \tag{4.9}$$

where U is characteristic flow speed, f is the Coriolis parameter, and L is a characteristic length scale. The relevant time scale is given by L/U which is the time it takes for flow to carry a parcel over a characteristic distance.

Calculations of Rossby numbers for previous exercises reveals that the Coriolis force could be neglected in most instances. The focus of the following exercises is placed on stratified processes that are influenced or even controlled by the Coriolis force.

4.1.4 Conservation of Potential Vorticity

Vorticity is the ability of flow to produce rotation. Frictionless flow that is almost in a geostrophic balance is called *quasi-geostrophic flow*. For a layered ocean, it can be shown that quasi-geostrophic flow conserves a quantity called *potential vorticity*. This statement is valid along flow trajectories and reads (Cushman-Roisin, 1994):

$$\frac{f + \xi_i}{h_i} = \text{constant} \tag{4.10}$$

where h_i is the thickness of a layer, and *relative vorticity* is given by:

$$\xi_i = \frac{\partial v_i}{\partial x} - \frac{\partial u_i}{\partial y} \tag{4.11}$$

In this context, the Coriolis parameter f is referred to as *planetary vorticity*. It is obvious from this principle that localised water-column stretching or squeezing can produce swift geostrophic flow disturbances.

4.1.5 Geostrophic Adjustment

Consider a situation of a surface layer of initial thickness h_1 and reduced density $\Delta\rho$ that occupies only a part of the domain in the x-direction but stretches to infinity in the y-direction (Fig. 4.2). The water column underneath this surface layer has an initial thickness of h_2. Obviously, this situation cannot persist at infinitum given the existence of a lateral pressure-gradient force that initially operates to draw ambient water under the lower-density surface layer. Convergence associated with this inflow creates a sea-level gradient that forces the surface layer in the positive x-direction. This lateral spreading, however, comes to a halt when the Coriolis force balances the lateral pressure gradient force. The result of this *geostrophic adjustment* is a *density front* (i.e. a zone of maximum density gradients) characterised by swift geostrophic flow along the frontal axis and little flow across.

On the basis of conservation of potential vorticity and the case of a relatively thin surface layer ($h_1 << h_2$), it can be shown that the width of the resultant frontal zone is given by the *internal Rossby radius of deformation*, defined by:

$$R = \frac{\sqrt{g'h_1}}{|f|} \tag{4.12}$$

It can be shown that the surface frontal geostrophic jet attains a speed of:

$$\left|v_{\text{geo}}\right| = \sqrt{g'h_1} \tag{4.13}$$

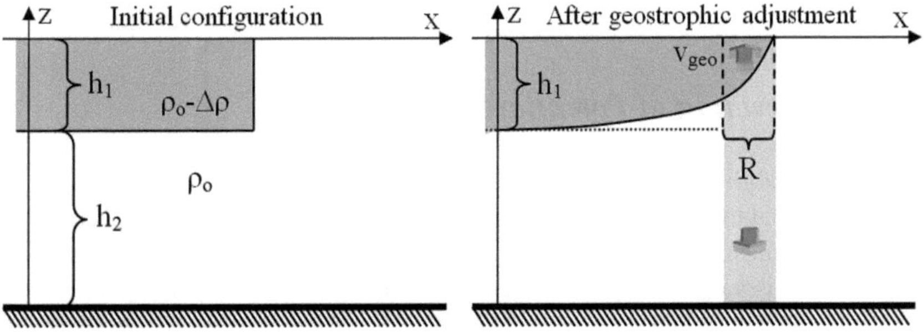

Fig. 4.2 Illustration of the geostrophic adjustment process

The speed of frontal flows decreases exponentially (on a spatial scale of R) with increasing distance from the surface outcrop of density surfaces. In the general case, a counterflow establishes in the water column underneath a surface density front. Density fronts in the coastal ocean attain typical widths of 1–20 km.

4.1.6 The 2.5d Shallow-Water Model

On the basis of vanishing gradients of variables in the y-direction and inclusion of the Coriolis force, the horizontal momentum equations for a vertical ocean slice can be written as:

$$\frac{\partial u}{\partial t} + u\frac{\partial u}{\partial x} + w\frac{\partial u}{\partial z} - fv = -\frac{1}{\rho_o}\frac{\partial(p+q)}{\partial x} + \text{Diff}(u) \tag{4.14}$$

$$\frac{\partial v}{\partial t} + u\frac{\partial v}{\partial x} + w\frac{\partial v}{\partial z} + fu = \text{Diff}(v) \tag{4.15}$$

where f is the Coriolis parameter and Diff(ψ) is the diffusion operator, being specified in Sect. 3.15. The other model equations are the same as in Exercise 12.

4.1.7 Implementation of the Coriolis Force

In the 2.5d version of the Arakawa C-grid, the v-components of velocity are calculated at pressure grid points (see Fig. 3.3). Hence, interpolation of velocity values is required for calculation of the Coriolis force. It can be shown that explicit formulation of the Coriolis force is numerically unstable. The nonhydrostatic solver of the Navier-Stokes equations is already formulated in an implicit manner in terms of dynamic pressure (see Sect. 3.4). Using another semi-implicit approach for the Coriolis force would lead to an even more complex solver. To avoid this, the Coriolis force is treated here by the local-rotation approach (described in Sect. 3.14 of Kämpf (2009)). With the sole presence of the Coriolis force, this approach leads to the finite-difference equations:

$$\begin{aligned} u^{n+1} &= \cos(\alpha)u^n + \sin(\alpha)v^n, \\ v^{n+1} &= \cos(\alpha)v^n - \sin(\alpha)u^n \end{aligned}$$

where the rotation angle is $\alpha = 2\arcsin(0.5\Delta t f)$. For sufficiently small numerical time steps of $\Delta t\,|f| << 1$, this can be approximated by $\alpha \approx \Delta t f$. With inclusion of other forces, this approach can be added to the first-guess of velocity components, yielding:

$$
\begin{aligned}
u^*_{i,k} &= \cos(\alpha)u^n_{i,k} + \sin(\alpha)v^n_u + \Delta t\ F^n_u \\
v^{n+1}_{i,k} &= \cos(\alpha)v^n_{i,k} - \sin(\alpha)u^n_v + \Delta t\ F^n_v \\
w^*_{i,k} &= w^n_{i,k} + \Delta t\ F^n_w
\end{aligned}
\tag{4.16}
$$

where $\alpha = \Delta t f$, u_v are u-values interpolated to v grid points, v_u are v-values interpolated to u grid points, and the remaining terms are given by:

$$
\begin{aligned}
F^n_u &= -\frac{1}{\rho_o \Delta x}(q^n_{i,k+1} - q^n_{i,k}) - \text{Adv}(u) + \text{Diff}(u) \\
F^n_v &= -\text{Adv}(v) + \text{Diff}(v) \\
F^n_w &= -\frac{1}{\rho_o \Delta z}(q^n_{i-1,k} - q^n_{i,k}) - \text{Adv}(w) + \text{Diff}(w)
\end{aligned}
\tag{4.17}
$$

where Adv(ψ) denotes the nonlinear terms. As before, the first-guess arrays u^* and w^* are input to the right-hand side of the Poisson equation (Eq. 3.25) to be solved by means of the S.O.R. iteration scheme.

4.1.8 Potential Problems

The step structure of the sea floor in z-coordinate models gives rise to a number of potential problems, summarised in the following.

- Each bottom step acts like an impermeable boundary and flows can only climb or descent these steps via creation of vertical velocity. For gravity currents cascading downward on the slope, this implies artificial situations of unstable density stratification.
- The condition of no flow through a vertical face of a bottom step requires use of zero-gradient conditions for dynamic pressure across this face. This, however, implies absence of geostrophic flow running parallel to a bottom step, which can substantially disturb the near-bottom geostrophic flow field.
- In the Arakawa C-grid, calculations of both the Coriolis force and bottom-stress terms require interpolation of surrounding velocity components (see Fig. 3.3). Discontinuities in vicinity of bottom steps lead to a bias in the dynamics.
- Vertical friction operates on horizontal section of the sea floor, whereas lateral friction occurs at vertical faces of bottom steps. The use of different values of horizontal and vertical turbulent viscosities creates dynamical irregularities. This can be avoided when disabling lateral momentum diffusion near bottom steps.

- Step structures of the sea floor are block-type obstacles for the vertically integrated continuity equation. This leads to a bias of the dynamics of gravity waves via partial reflection.

These shortcoming, arising when using z-coordinate models, are unfortunate and can only be avoided with the choice of other coordinate systems (such as sigma-coordinate models) or model types (such as layer models).

4.2 Exercise 16: Geostrophic Adjustment

4.2.1 Aim

This exercise employs the 2.5d vertical ocean-slice model to study geostrophic adjustment of a surface density front of infinite length.

4.2.2 Task Description

We consider a model domain, 50 km in length and 500 m in depth, resolved by a lateral grid spacing of $\Delta x = 500$ m and a vertical grid spacing of $\Delta z = 10$ m. Lateral boundaries are closed. An isolated surface layer of 10 km in width and 250 m in thickness is initially placed in the centre of the model domain (Fig. 4.3). This layer is lighter compared with the ambient ocean in which the density is $\rho = 1{,}028\ \text{kg/m}^3$. The density anomaly of the surface layer is linearly adjusted from zero to a final value of 0.1 kg/m^3 over the first 6 hrs of the simulation. No further forcing is applied afterward and the dynamics can evolve freely.

Small uniform isotropic values of $1\times10^{-4}\ \text{m}^2/\text{s}$ are used for both eddy diffusivity and eddy viscosity. The bottom friction parameter is set to $r = 0.001$. The total simulation time is 60 hrs with hourly data outputs. The time step is set to $\Delta t = 5$ s. The model includes a freely moving sea surface. The pressure accuracy of the S.O.R. scheme is set to $\epsilon = 0.01$ Pa. Two scenarios are considered. Rotational effects are ignored in the first scenario by setting the Coriolis parameter to zero. The second scenario uses a mid-latitude value of the Coriolis parameter of $f = 1 \times 10^{-4}\ \text{s}^{-1}$ (Northern Hemisphere).

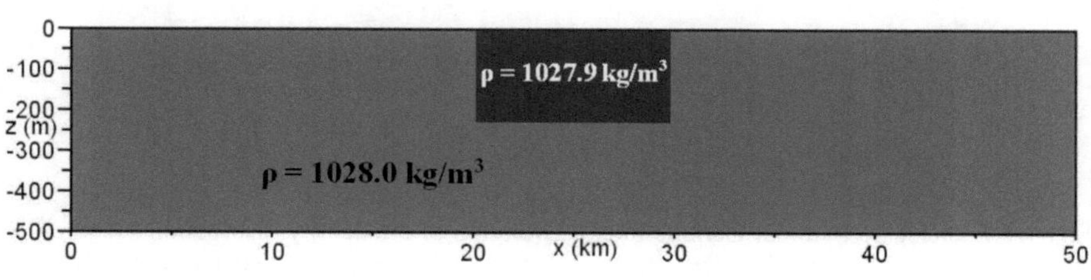

Fig. 4.3 Initial configuration for Exercise 16

4.2.3 Results

Without the Coriolis force, the low-density surface layer spreads laterally towards the ends of the channel (Fig. 4.4). According to theory (Eq. 3.48), the density fronts should propagate at a speed of 50 cm/s, which agrees reasonably well with the simulation. Frontal flows become reflected at the closed lateral boundaries and meet again in the centre of the model domain after 50 min. As a consequence of volume conservation, initial outward spreading of the surface layer is compensated by inward currents in the water column underneath. In a steady state, the low-surface density layer will cover the entire domain and the currents will eventually come to rest.

Circular ocean eddies are largely geostrophic. Although the 2.5d ocean slice model does not capture three-dimensional features, results shown in the following are analog to those of a vertical section cutting through the centre of a symmetrical ocean eddy. Accordingly, the patterns discussed below can be interpreted as cyclonic or anticyclonic vortices. A cyclonic eddy has a low-pressure centre, whereas an anticyclonic eddy is characterised by a high-pressure centre.

The initial lateral spreading of the density front ceases under influence of the Coriolis force. Instead of continued spreading, lateral pressure gradients create geostrophic flow that run parallel to the density fronts and not across. As a consequence of this, the low-density surface layer remains confined in horizontal extent (Fig. 4.5). The simulated width of the frontal zone agrees well with theory noting that Eq. (4.12) predicts a value of 4.9 km. The frontal flow attains speeds of

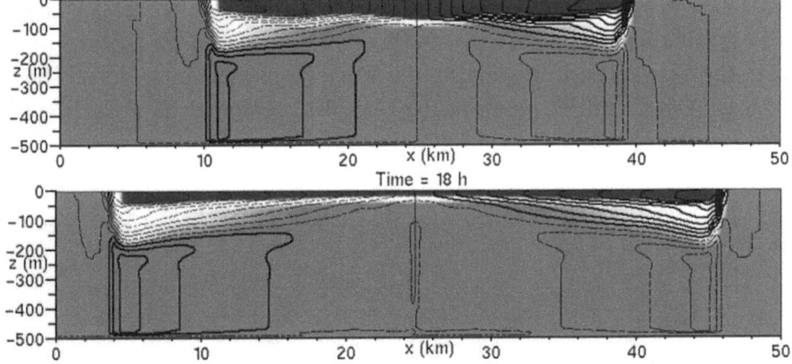

Fig. 4.4 Exercise 16 (without Coriolis force). Evolution of the density distribution (*shading*) at selected times of the simulation. *Lines* are contours of u with a contour interval of 0.05 m/s. *Solid* (*broken*) *lines* denote positive (negative) speeds

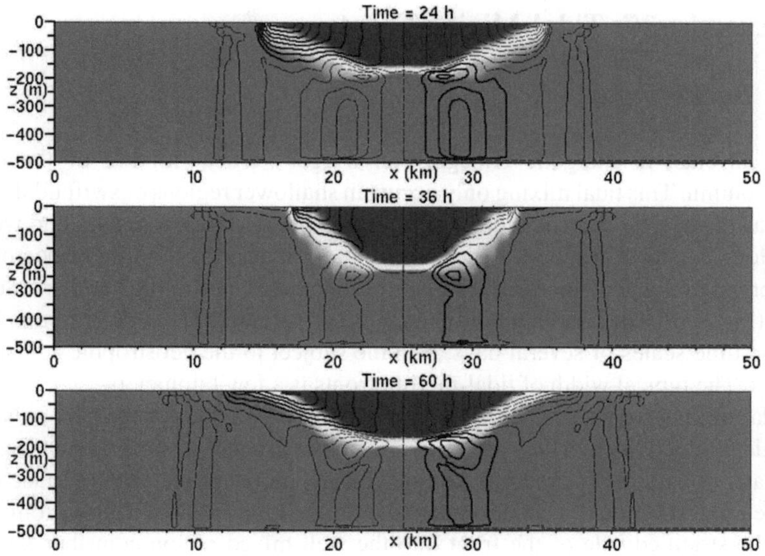

Fig. 4.5 Exercise 16 (with Coriolis force). Evolution of the density distribution (*shading*) at selected times of the simulation. *Lines* are contours of v with a contour interval of 0.05 m/s. *Solid* (*broken*) *lines* denote positive (negative) speeds

0.35 ± 15 cm/s. The time-averaged flow speed is slightly lower that the theoretical value of 49 cm/s, derived from Eq. (4.13), presumably because of artificial lateral diffusive smoothing of steep gradients in the sea level.

Interestingly, initial water-column stretching creates deep flows running opposite to those establishing in the surface layer. This feature is important to remember for studies of rotational exchange flows through oceanic straits, to be investigated in Exercise 22.

It is obvious that temporal variations of flow speeds are the signature of inertial oscillations. These oscillations lead to an oscillatory pattern alternating between lateral stretching and shrinking of the low-density surface lens. The Coriolis parameter used corresponds to an inertial period is 17.5 hrs, which agrees well with the prediction (not shown). The reader is encouraged to verify the latter statement. Overall, the 2.5d vertical ocean-slice model appears capable of successfully simulating rotational effects incurred by the Coriolis force.

4.2.4 Additional Exercise for the Reader

Repeat this exercise, but place an isolated layer of denser water on the sea floor. Explore the resultant evolution of the density field for different values (including zero) of the bottom friction parameter.

4.3 Exercise 17: Tidal-Mixing Fronts

4.3.1 Background

Tidal currents, if energetic enough, can trigger vertical mixing of the entire water column. This tidal mixing only occurs in shallower regions of swift tidal flows, whereas the density stratification in adjacent deeper portions of the sea and weaker tidal flows remains stratified. A density front establishes as a consequence of this differential mixing and marks the transition zone between stratified and unstratified water (Fig. 4.6). The resultant horizontal pressure gradient force gives rise to flows that, on time scales of several days, become subject to the geostrophic adjustment process. The typical width of tidal-mixing fronts is a few kilometers.

Tidal-mixing fronts, such as those in the Irish Sea, are known for their enhanced biologic productivity and they are generally rich in fish and seabird abundance. Bottom water of the stratified regime contain typically more nutrients than surface water and the mixed regime. It is believed that the flux of high-nutrient near-bottom water from the stratified side of the front into the well-mixed region contributes to this (Mann and Lazier, 1996). Most tidal mixing fronts only exist during the warmer seasons of the year given that solar heating is required for establishment of thermal stratification of the water column. Tidal mixing fronts are usually found closer to the shore during spring tides than during neap tides.

The ratio between local water depth h and the cube of the average speed U of tidal flows is often used as an indicator of the location of a tidal-mixing front (Mann and Lazier, 1996). Advanced methods consider effects of intensified wind-induced mixing and air-sea heat fluxes. A discussion thereof is beyond the scope of this book.

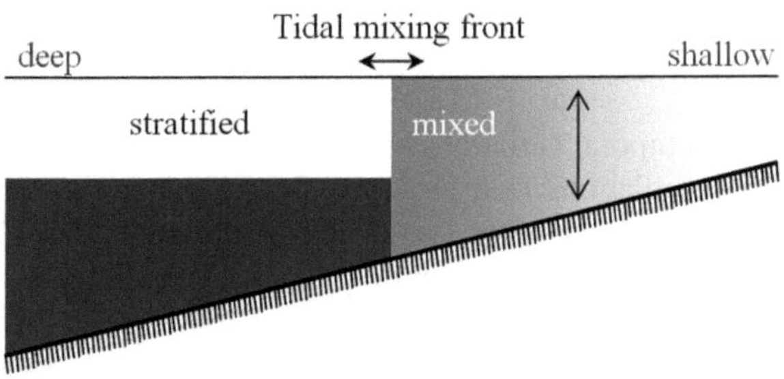

Fig. 4.6 Schematic of the density distribution created by enhanced tidal mixing in shallower water

4.3.2 Task Description

The model domain is 10 km in length. Total water depth linearly decreases from 100 m at one side to 50 m at the other side of the domain (Fig. 4.7). Lateral boundaries are closed. The horizontal grid spacing is set to $\Delta x = 100$ m and a vertical

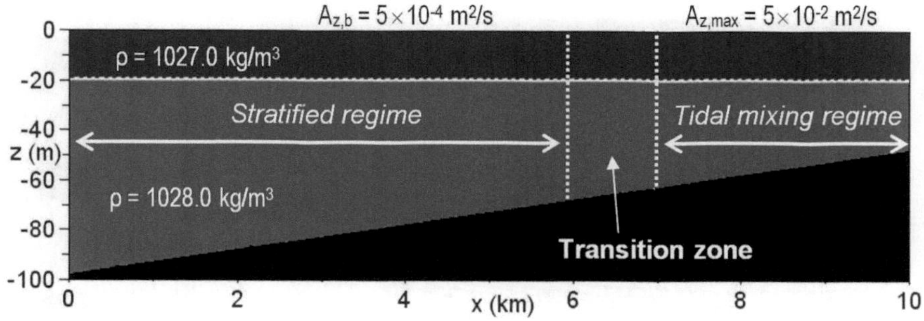

Fig. 4.7 Initial configuration for Exercise 17

grid spacing of $\Delta z = 2\,\mathrm{m}$ is used. Initially, the water column is stably stratified. The surface mixed layer is 20 m thick and has a density of $\rho = 1{,}027\,\mathrm{kg/m^3}$. The density of seawater underneath is set to $\rho = 1{,}028\,\mathrm{kg/m^3}$. Density fluxes across the sea surface are neglected.

Instead of attempting to simulate tidal currents and the turbulence associated with this, this exercise treats the creation of tidal-mixing fronts in a simplified manner, as described in the following. Background values of vertical eddy viscosity and eddy diffusivity of $A_{z,b} = K_{z,b} = 5 \times 10^{-4}\,\mathrm{m^2/s}$ are prescribed in the entire model domain. The effect of tidal mixing in shallower water is parameterised via prescription of increased turbulence levels. To this end, enhanced turbulence levels of maximum values of $A_{z,\max} = K_{z,\max} = 5 \times 10^{-2}\,\mathrm{m^2/s}$ are prescribed in regions shallower than 60 m. To mimic the tidal cycle, the latter is modulated on a semi-diurnal period according to:

$$A_z = K_z = A_{z,\max}\,|\sin(2\pi t/T)| + A_{z,b}$$

where t is time, and T is the forcing period. Linear interpolation of turbulent levels is applied in a transition zone of 1 km in width. Horizontal eddy viscosity and eddy viscosity are kept at a small uniform value of $A_h = K_h = 1 \times 10^{-4}\,\mathrm{m^2/s}$. Quadratic bottom friction is included with $r = 0.001$. The Coriolis parameter is set to $f = 1 \times 10^{-4}\,\mathrm{s^{-1}}$ (mid-latitudes of the Northern Hemisphere). The total simulation time is 24 hrs with data outputs at 15-min interval. The time step is set to $\Delta t = 1.5\,\mathrm{s}$, using the free-surface version of the 2.5d vertical ocean-slice model. Pressure accuracy of the S.O.R. simulation is set to $\epsilon = 0.001\,\mathrm{Pa}$.

4.3.3 Results

Enhanced tidal stirring mixes the entire water column in shallower water whereas density stratification remains in deeper regions of the model domain (Fig. 4.8). Surface water of the mixed regime is denser compared with that in the nearby stratified

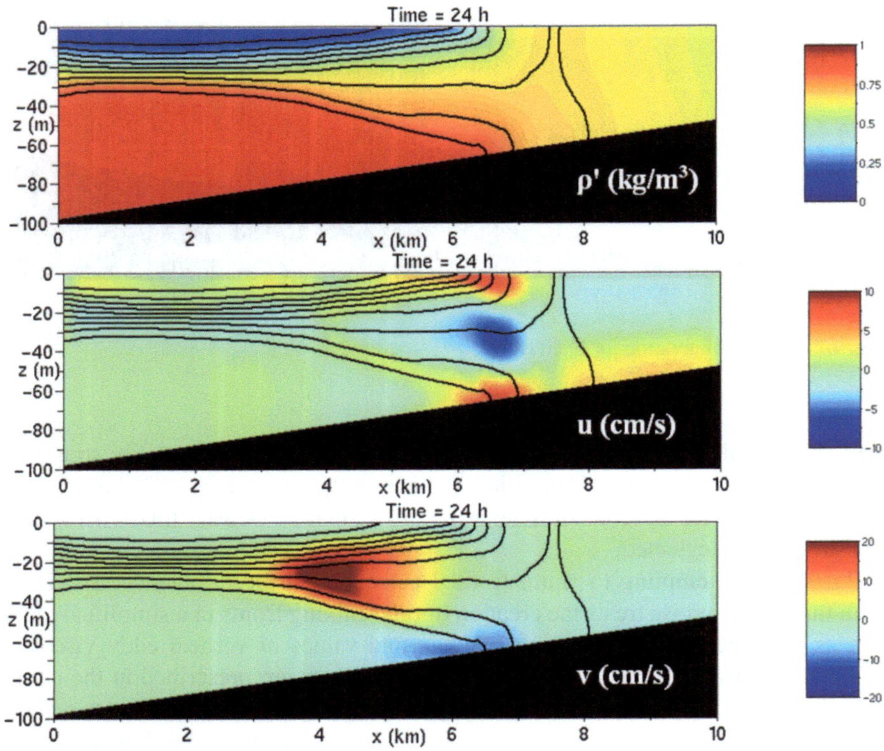

Fig. 4.8 Exercise 17: Distributions of density anomaly relative to $\rho_o = 1{,}027\,\mathrm{kg\,m^{-3}}$ (*top panel*), u-component of velocity (*middle panel*), and v-component of velocity (*bottom panel*) after 24 hrs of simulation

regime. Bottom water of the mixed regime is less dense than adjacent water of the stratified regime. The resultant horizontal pressure gradients give rise to a three-layer circulation. This consists of onshore flows in both surface and bottom layers and offshore flow in the middle layer, the latter injecting water from the mixed regime into the pycnocline of the stratified regime (see Fig. 4.8). Lateral injection of water into the pycnocline continues throughout the simulation at speeds varying between 5 and 10 cm/s. These onshore and offshore flows are directly driven by lateral pressure gradients and, therefore, constitute *ageostrophic* components of the circulation. The ageostrophic cross-shelf circulation intensifies during weak tidal flows (parameterised in the model) and weakens during times of enhanced ambient vertical mixing.

Geostrophic adjustment leads to a swift subsurface frontal geostrophic jet of 20 cm/s in speed in the middle layer. The direction of this flow is such that the shallower water is located on its right-hand side in the Northern Hemisphere. Weaker geostrophic flows of the opposite direction establish near both the surface and the bottom. The resultant frontal zone has an approximate width of 3 km. In the real situation, geostrophic jets of tidal mixing fronts (and other density fronts) tend to become dynamically unstable and break up into mesoscale eddies inducing vigorous lateral mixing across the front. This cannot be simulated with the 2.5d model.

4.3.4 Additional Study

Prognostic advection-diffusion equations for Eulerian tracer concentration fields are added to the code to quantify the sources of water that make up the frontal zone. To this end, the author decided to use three separate tracer fields (Fig. 4.9). One field is allocated an initial concentration of 100% in the well-mixed regime and zero values elsewhere. The other two tracer fields mark the surface and bottom layers of the stratified regime.

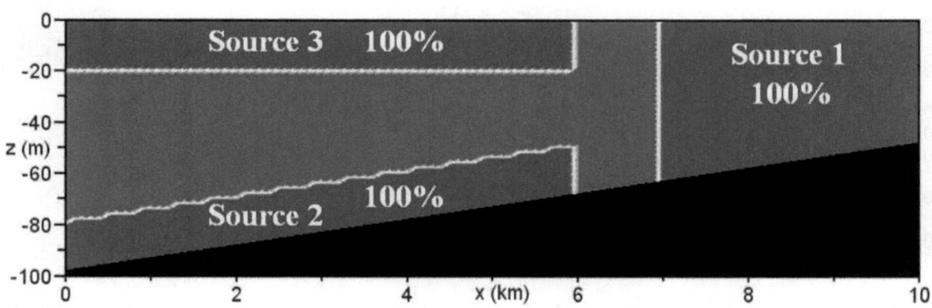

Fig. 4.9 Exercise 17: Initial distribution of three separate Eulerian tracer fields

4.3.5 Results and Discussion

After 24 hrs of simulation, the stratified regime has contributed 19% of near-bottom water and 13% of near-surface water to the frontal zone (Fig. 4.10). The well-mixed regime outside the front has contributed 23% to the frontal zone. The remainder 45% of water stems from other sources. The timescale of semi-diurnal tidal variations, mimicked here via variation of turbulence levels, is shorter than the inertial period (which is about 17.5 hrs for the model configuration), so that the geostrophic adjustment process remains incomplete. Instead of this, each period of enhanced tidal stirring is followed by a phase of gravitational adjustment. During this process, surface and bottom water of the stratified regime are displaced a few kilometers closer to the shore and previously mixed water becomes drawn into the pycnocline of the stratified regime. The result of this gravitational adjustment is a pumping of both surface and bottom water from the stratified regime into the frontal zone. Convergence-induced upwelling of near-bottom water supports this process (see middle panel in Fig. 4.10). Findings, shown here, confirm the injection of nutrient

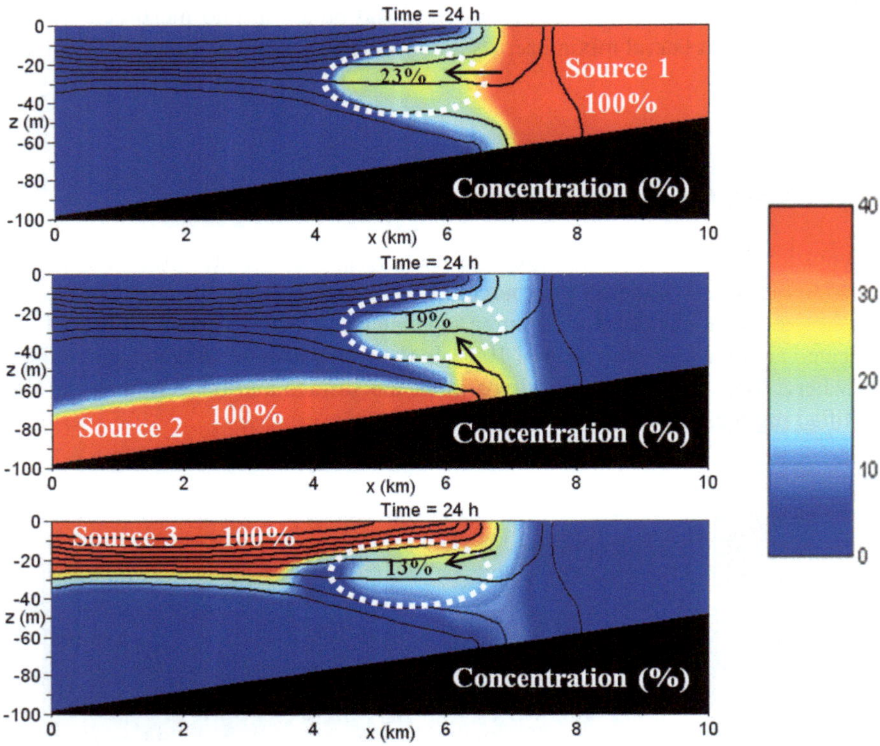

Fig. 4.10 Exercise 17: Distributions of the three Eulerian tracer fields after 24 hrs of simulation

enriched bottom water from the stratified side of the front into the frontal zone as stimulus of biologic production.

4.3.6 Additional Exercises for the Reader

Repeat this exercise with an initial density difference between the top and the bottom layers of $\Delta\rho = 2\,\mathrm{kg\,m^{-3}}$, which is twice that used before, and explore the features of the resultant tidal mixing front and associated flows.

4.4 Coastal Upwelling

4.4.1 Background

Coastal upwelling is a process by which colder and often nutrient-enriched water is lifted closer the sea surface where increased light intensity promotes the production of phytoplankton – a key trigger of the marine food chain.

4.4.2 How Does It Work?

Coastal upwelling is created by winds blowing along the coast such that the surface Ekman drift is directed offshore. This offshore movement lowers the coastal sea level and produces a coastal geostrophic jet running into the same direction as the wind. Friction of this flow with the seabed creates onshore flow in the bottom Ekman layer and upward movement occurs in vicinity of the coast. The upward tilt of density surfaces near the coast, however, reduces lateral pressure gradients in the bottom layer. This partial compensation of the imposed surface pressure field in the ocean interior, called *baroclinic compensation* and sometimes referred to as *buoyancy shutdown* (Chapman, 2002), weakens the geostrophic flow near the bottom and thus the onshore flow in the bottom Ekman layer.

4.4.3 Partial and Full Upwelling

Consider a coastal ocean consisting of two superimposed layers of different densities. Full upwelling refers to a situation in which the density interface has reached the surface and forms a density front that is displaced offshore leaving upwelled cold water exposed at the surface (Fig. 4.11). Partial upwelling occurs for a brief or weak wind event such that the interface has upwelled but not to the point of reaching the surface.

The *reduced-gravity concept* for a two-layer fluid is based on the assumption that the density interface between the layers adjusts such to sea-level gradients that horizontal pressure gradients and hence the flow vanishe in the bottom layer. The reduced-gravity concept implies that:

$$P_2 = 0 = \rho_1\, g\, \eta_1 + (\rho_2 - \rho_1)\, g\, \eta_2$$

where ρ_1 and ρ_2 are densities of the top and bottom layers, η_1 is surface elevation, and η_2 is the elevation of the density interface. This leads to a relation between sea-level elevations and interface displacements according to:

$$\eta_1 = -\frac{\rho_2 - \rho_1}{\rho_1}\eta_2$$

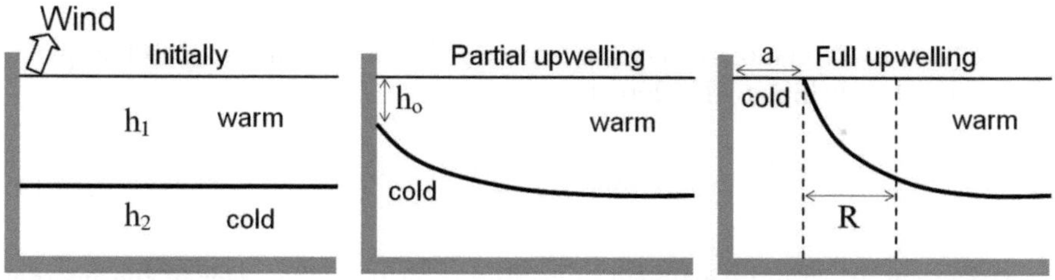

Fig. 4.11 Illustration of the coastal upwelling process (Northern Hemisphere). Adapted from Cushman-Roisin (1994)

Oceanographers frequently use this relation to estimate slopes of the sea level, driving the surface geostrophic flow, from the observed slope of the pycnocline. On the other hand, the thickness of the surface mixed layer is given by:

$$h = h_o + \eta_1 - \eta_2$$

where h_o is the undisturbed thickness. Given that the magnitude η_2 typically exceeds that of η_1 by far, the barotropic pressure gradient can be formulated according to the reduced-gravity concept as:

$$-g\frac{\partial \eta_1}{\partial x} = -g'\frac{\partial h}{\partial x}$$

where reduced gravity is defined by $g' = (\rho_2 - \rho_1)/\rho_2 g$. Using the reduced-gravity concept for a two-layer ocean, the offshore distance a of the outcrop of the density interface for full upwelling can be estimated from (see Cushman-Roisin, 1994):

$$a = \frac{I}{|f|} - R \tag{4.18}$$

where $R = \sqrt{g'h_1}/|f|$ is the internal Rossby radius of deformation, and the so-called "wind impulse" is defined by:

$$I = \frac{1}{\rho_o h_1}\int_{\text{event}} \tau \, dt$$

with τ being the alongshore component of upwelling favorable wind stress. When averaging this wind-stress component over a certain time span t^*, the latter equation can be expressed as:

$$I = \frac{\langle \tau \rangle}{\rho_o h_1} t^*$$

According to Eq. (4.18), the transition between partial and full upwelling occurs when:

$$I = |f| R$$

The latter two equations can be combined to yield an estimate of the time span it takes for full upwelling to develop; that is,

$$t^* = \rho_o h_1 \frac{\sqrt{g'h_1}}{\langle \tau \rangle} \tag{4.19}$$

The width of a fully developed upwelling front is of the order of the internal radius of deformation R (Cushman-Roisin, 1994).

4.4.4 The Upwelling Index

The *upwelling index* can be used as an indicator and measure of coastal upwelling events. This index is an estimate of the magnitude of offshore transport in the surface Ekman layer and it is calculated from:

$$\text{Upwelling index} = \frac{\left|\tau^{\text{wind}}\right| \cos\left(\alpha^*\right)}{|f|\,\rho_o} \tag{4.20}$$

where τ^{wind} is the wind-stress magnitude, and α^* is the relative angle between wind direction and coastline orientation. The upwelling index carries units of square metres per second, which refers to a volume transport in cubic metres per second per unit length of the coastline. In contrast to coastal upwelling, *coastal downwelling* refers to a situation in which the Ekman drift pushes surface water against the coast and which corresponds to negative values of the upwelling index.

4.5 Exercise 18: Coastal Upwelling and Downwelling

4.5.1 Aim

The aim of this study is to explore the wind-driven coastal upwelling process in a stratified water column with the 2.5d vertical-ocean slice model.

4.5.2 Task Description

The model domain is 50 km in length, resolved by a horizontal grid spacing of Δx = 500 m, and has a maximum depth of 100 m, resolved by a vertical grid spacing of Δz = 5 m (Fig. 4.12). The water depth decreases from 100 to 50 m along the length of the model domain. Shallow water is bounded by a coast. The offshore lateral boundary is treated as an open boundary using zero-gradient conditions for all variables except for sea-level elevation. To avoid that the model domain empties

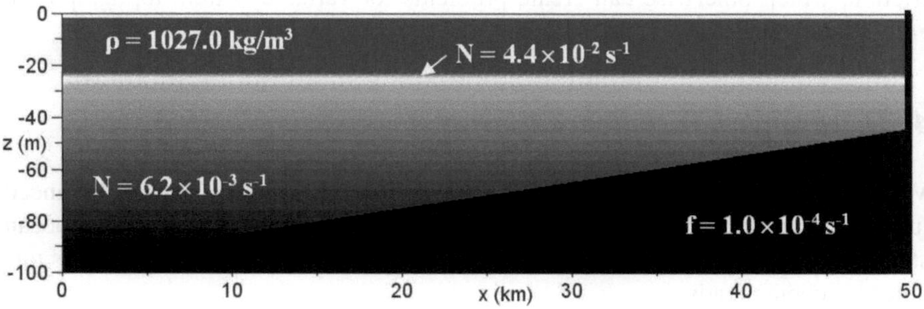

Fig. 4.12 Initial configuration for Exercise 18

over time under continuous offshore Ekman transport, the sea level at the open boundary is kept at its initial value throughout the simulation.

The surface-mixed layer is initially 25 m thick and of a density of 1,027 kg/m^3. A pycnocline is situated at the base of this layer across which density changes by 1 kg/m^3. This corresponds to a local stability frequency of $N = 4.4 \times 10^{-2}\,\mathrm{s}^{-1}$. Density is assumed to increase with depth underneath the pycnocline with a stability frequency of $N = 6.2 \times 10^{-3}\,\mathrm{s}^{-1}$. The Coriolis parameter is set to $f = -1 \times 10^{-4}\,\mathrm{s}^{-1}$, representing mid-latitudes in the Southern Hemisphere. Recall that, in the Southern Hemisphere, the net wind-driven water movement in the surface Ekman layer is at right angle and to left with respect to the wind direction.

Horizontal eddy viscosity and eddy viscosity are set to a constant value of $A_h = K_h = 1\,\mathrm{m}^2/\mathrm{s}$. Vertical eddy viscosity and diffusivity, assumed to be equal, are diagnosed from an advanced turbulence closure scheme, described in the next section. A value of $K_z = A_z = 0.05\,\mathrm{m}^2/\mathrm{s}$ is prescribed near the surface to mimic the immediate effect of wind mixing. Ambient background values of $A_{z,\min} = K_{z,\min} = 1 \times 10^{-4}\,\mathrm{m}^2/\mathrm{s}$ are used. The bottom friction parameter (assuming a quadratic bottom-friction law) is set to $r = 0.003$.

The model is forced via prescription of a alongshore wind stress of the form:

$$\tau_y^{\mathrm{wind}} = \tau_o \sin(2\pi t/T) \tag{4.21}$$

where t is time, and the period T is chosen as 10 days, being characteristic for synoptic weather events. Wind-stress components are defined at the same locations as u and v.

Two different scenarios are considered, both running over a total of 5 days with data output every 3 hrs. The first scenario uses $\tau_o = +\,0.2\,\mathrm{Pa}$ corresponding to upwelling favorable coastal winds, whereas the second scenario uses $\tau_o = -0.2\,\mathrm{Pa}$, which should lead to coastal downwelling. The mean wind-stress magnitude over the 5-day simulation period is $2\,|\tau_o|\,/\pi = 0.128\,\mathrm{Pa}$. The surface stress felt by the ocean is calculated from Eq. (2.5).

This application uses a free surface and a numerical time step of $\Delta t = 12\,\mathrm{s}$. The pressure accuracy of the S.O.R. iteration is set to $\epsilon = 1 \times 10^{-4}\,\mathrm{Pa}$. Zero-gradient conditions are employed for the alongshore velocity component v near "dry" grid cells. This condition, called *full-slip boundary condition*, disables lateral friction, which otherwise can create problems for variable bottom topography in z-coordinate models.

4.5.3 Advanced Turbulence Closure

Vertical eddy diffusivity and eddy viscosity can be diagnosed from an advanced turbulence closure proposed by Kochergin (1987). When applied to a vertical ocean slice and under the assumption that eddy viscosity equals eddy diffusivity, this turbulence scheme reads:

$$K_z = A_z = (c_2 \Delta z)^2 \sqrt{(\partial u/\partial z)^2 + (\partial v/\partial z)^2 - N^2} + A_{z,\min} \tag{4.22}$$

where c_2 is a free parameter, Δz is vertical grid spacing, and $A_{z,\min}$ is a small background value. The author uses $c_2 = 0.15$. Additional treatment is required if the grid spacing chosen does not resolve convective plumes created by unstable density stratification ($N^2 < 0$). A conventional approach is to increase turbulence levels to a certain large value $A_{z,\max}$ in such situations. Explicit finite-difference formulation of the diffusion terms is associated the stability condition:

$$\Delta t \leq \frac{(\Delta z)^2}{\max(A_z)}$$

4.5.4 Results: Upwelling Scenario

Full upwelling establishes in response to the wind forcing applied (Fig. 4.13). The wind forcing produces offshore Ekman drift in the surface layer and onshore Ekman drift in the bottom layer. Speeds these Ekman-layer currents are 10–15 cm/s. Currents in the bottom Ekman layer move deeper water toward the coast and upward

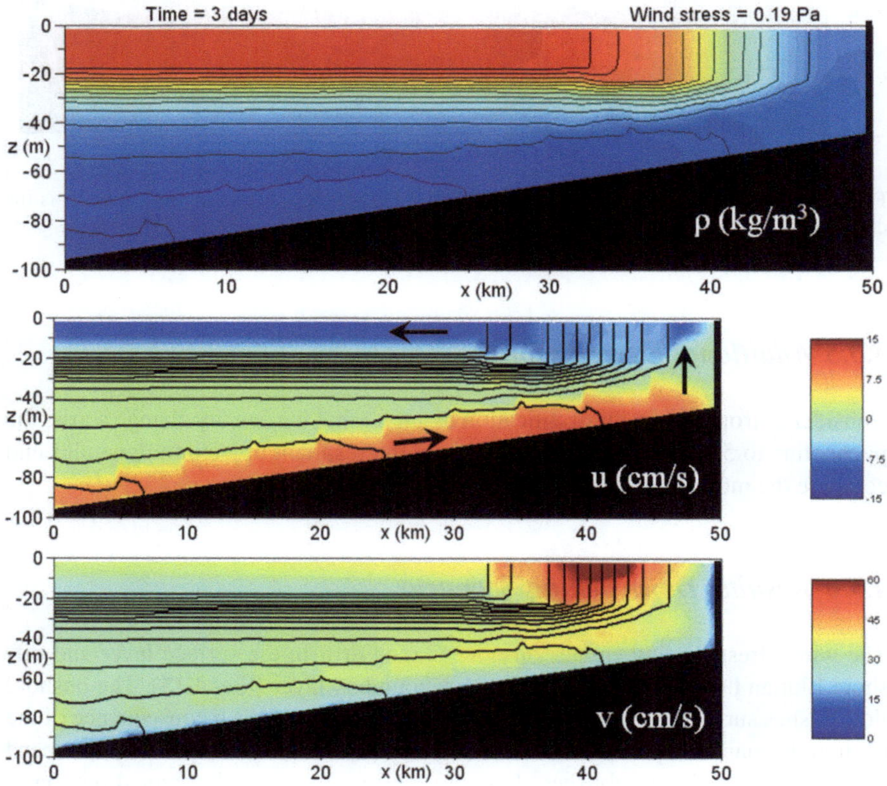

Fig. 4.13 Exercise 18. Upwelling scenario. Snapshots of the density field, and cross-shore and along shore components of horizontal velocity after 3 days of simulation

to the surface. The bottom Ekman layer is the result of geostrophic alongshore flow that runs into the same direction as the wind. This geostrophic flow is magnified in vicinity of the surface density front, created by the upwelling, where it attains speeds of 60 cm/s. Step-type representation of variable bathymetry obviously induces some artificial local disturbances, but this does not lead to numerical instabilities.

Theory gives a frontal speed of $c = \sqrt{(g'h_1)}$ of 49 cm/s and an internal Rossby radius of deformation $R = c/|f|$ of 4.8 km. Equation (4.18) suggests that the distance of the density outcrop from the coast is a = 11.8 km after 5 days of simulation. Equation (4.19) suggests that the density interface outcrops at the surface after 1.14 days. The model predictions are in good agreement with theory (see Fig. 4.14). The coastal sea level drops by 20 cm (not shown) as a consequence of offshore Ekman drift. The magnification of the geostrophic flow in vicinity of the surface density front is caused by a steepening of the sea-level gradient in this region.

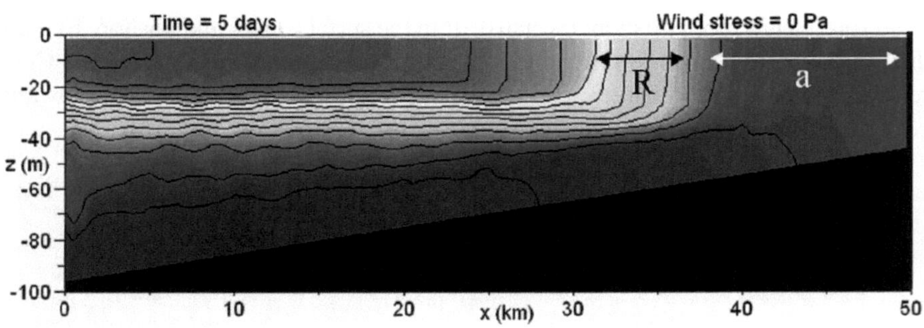

Fig. 4.14 Exercise 18. Same as *top panel* of Fig. 4.13, but after 5 days of simulation. a is the distance of the density outcrop from the coast. R is the frontal width

4.5.5 Additional Exercise for the Reader

Consider a stronger density stratification by increasing the density change across the pycnocline to 5 kg/m^3. Explore variations in the resultant upwelling dynamics and compare the model prediction with theory.

4.5.6 Results: Downwelling Scenario

The wind-stress forcing creates onshore Ekman drift in the surface layer and offshore Ekman flow in the frictional bottom-boundary layer (Fig. 4.15). The onshore flow pushes surface water against the coast and downward. As a consequence of the onshore Ekman transport in the surface layer, a geostrophic jet of 50 cm/s in speed and a width of 10 km establishes along the coast running into the same direction as the wind. This alongshore geostrophic flow triggers net offshore drift in the bottom

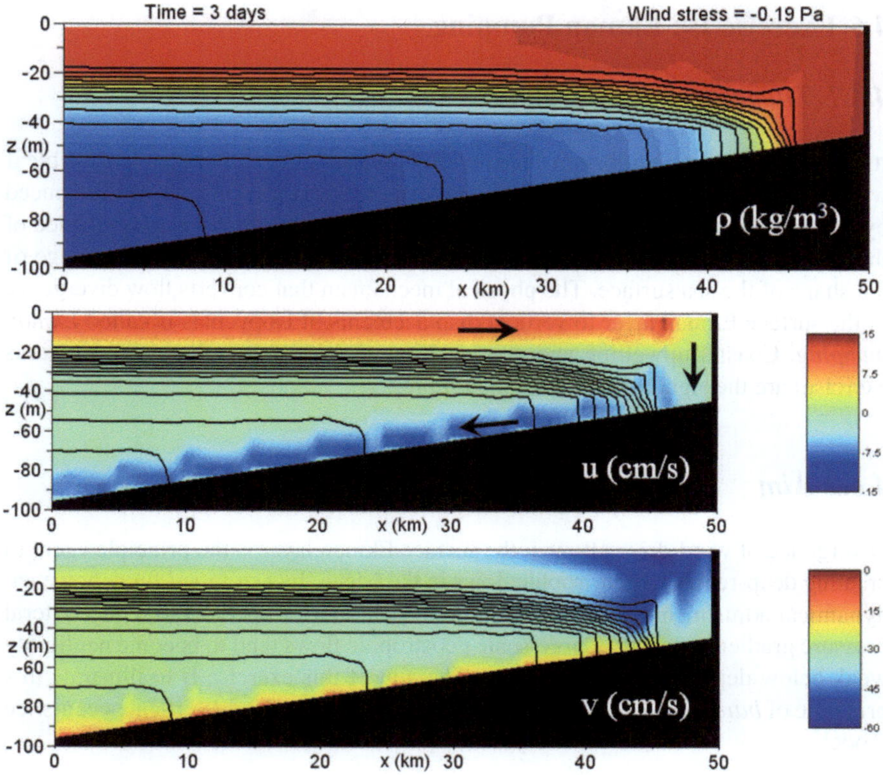

Fig. 4.15 Exercise 18. Downwelling scenario. Snapshots of the density field, and cross-shore and alongshore components of horizontal velocity after 3 days of simulation

Ekman layer. Notice that the bottom Ekman layer attains a thickness of 10–20 m in this simulation. During the course of the downwelling event, the coastal sea level gradually rises by 20 cm (not shown). The resultant barotropic offshore pressure gradient is the principal driver of the geostrophic alongshore jet.

4.5.7 Additional Exercise for the Reader

Explore the density structure and flows resulting from oscillatory wind-stress forcing, prescribed by Eq. (4.21), with a period of 4 days for a total simulation time of 20 days. Note that the average wind stress is zero, so that, without diffusion effects, the final density distribution should be the same as the initial distribution. Over time, however, we anticipate that diffusion causes residual lateral density gradients and associated geostrophic flows.

4.6 Exercise 19: Ekman Pumping

4.6.1 Theoretical Background

It is obvious that an equilibrium distribution of sea-level elevation is only possible if wind-induced flow divergence (convergence) in the surface Ekman layer is balanced by a flow convergence (divergence) in the ocean underneath. As a consequence of this, density surfaces in the ocean interior tend to be an amplified mirror image of the shape of the sea surface. The physical mechanism that converts flow divergence in the surface Ekman layer to vertical displacements of isopycnals is called *Ekman pumping*. Coastal upwelling and downwelling of density interfaces (see previous exercise) are the signatures of Ekman pumping.

4.6.2 Aim

Divergence of wind-driven flow in the surface Ekman layer is the principle agent to creating deep-reaching geostrophic flows in the ocean. In a stratified fluid, however, dynamical adjustment of isopycnals in the ocean interior operates to reduce lateral pressure gradients such that large-scale geostrophic flows tend to become negligibly weak below depths of 1,500–2,500 m. The aim of this exercise is to illustrate this principle of *baroclinic compensation* using the wind-forced 2.5d vertical ocean-slice model.

4.6.3 Task Description

The model domain has a length of 500 km, resolved by a horizontal grid spacing of $\Delta x = 5$ km, and a depth of 500 m, resolved by a vertical grid spacing of $\Delta z = 20$ m (Fig. 4.16). The unrealistically small depth has been chosen to allow for relatively long numerical time steps. Initially, the ocean is at rest and void of lateral density variations. The surface mixed-layer is 200 m thick and has a density of $\rho = 1{,}025$ kg/m^3. There is a density change at the base of this layer across which density changes by $\Delta\rho = 5$ kg/m^3. The associated stability frequency of this pycnocline

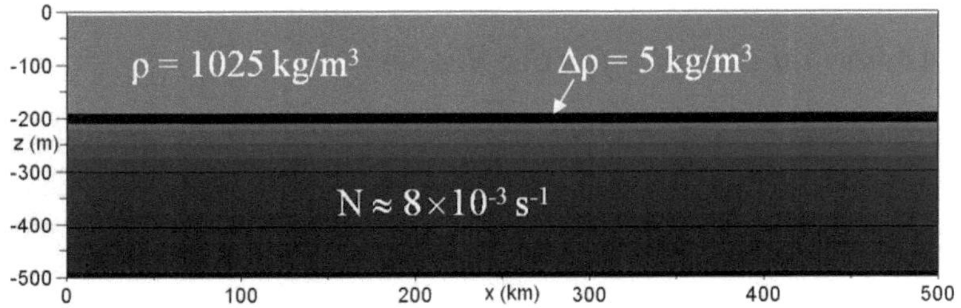

Fig. 4.16 Initial configuration for Exercise 19

is $N \approx 0.05\,\mathrm{s}^{-1}$. Below the pycnocline, density increases linearly with depth characterised by a stability frequency of $N \approx 8 \times 10^{-3}\,\mathrm{s}^{-1}$. The Coriolis parameter is set to a constant value of $f = 1 \times 10^{-4}\,\mathrm{s}^{-1}$ (Northern Hemisphere). Variation of the Coriolis parameter with geographical latitude is ignored.

The model is forced by prescription of wind stress acting in the y-direction and thus normal to the model slice. Three different wind forcings are considered (Fig. 4.17). The structure of the wind-stress forcing is:

$$\tau_y^{wind} = \tau_o \cos(\pi x/L) + \tau_1 \qquad (4.23)$$

where L is the horizontal extent of the model domain. The first scenario uses $\tau_o = -0.1$ Pa and $\tau_1 = 0$, the second scenario $\tau_o = +0.1$ Pa and $\tau_1 = 0$, and the third scenario $\tau_o = +0.05$ Pa and $\tau_1 = +0.05$ Pa. In each case, the wind field is gradually adjusted from zero to its final values during the initial 2 days of simulation to avoid unwanted initial disturbances in the form of gravity waves and inertial oscillations.

The wind-stress forcing is such that surface Ekman transports are free of lateral divergence at lateral boundaries. This implies that the sea-level elevation, the dynamic pressure part q and vertical velocity should remain at zero values at these boundaries. This is consistent with choice of zero-gradient conditions for u, which, in turn, specify the indices of boundary grid cells (see Fig. 4.18). Zero-gradient lateral boundary conditions are used for the remainder variables; that is, the velocity component normal to the model slice v and density.

Uniform values of $A_h = K_h = 50\,\mathrm{m}^2/\mathrm{s}$ are used for horizontal diffusivity and viscosity. The Kochergin turbulence scheme of previous exercises is used to calculate

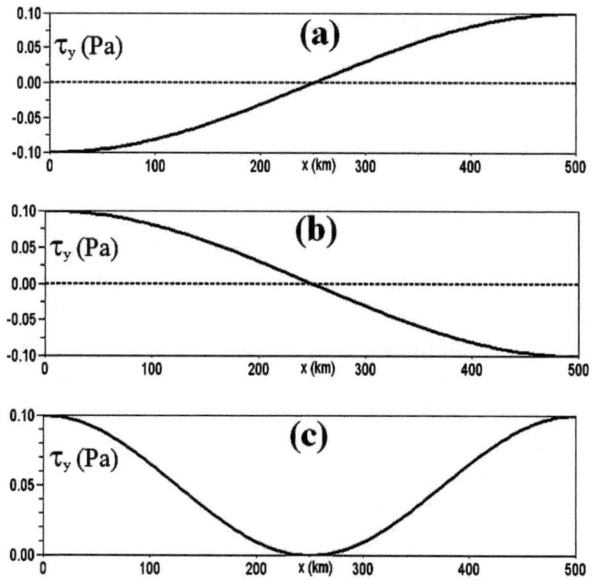

Fig. 4.17 Three different steady wind-stress forcings for Exercise 19

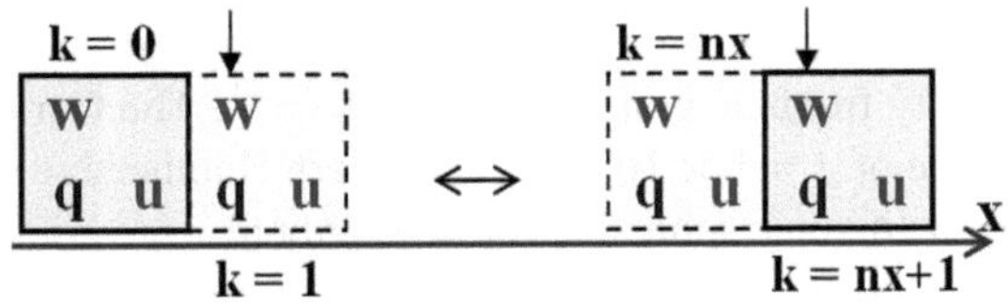

Fig. 4.18 Illustration for implementation of lateral boundary conditions. *Vertical arrows* indicate the grid columns in which values of vertical velocity and of the nonhydrostatic part of dynamic pressure q are set to zero

values of vertical eddy viscosity and diffusivity. The bottom friction parameter is set to $r = 0.001$. The total simulation time is 50 days with half daily data outputs. The pressure accuracy for the S.O.R. iteration is set to $\epsilon = 10^{-3}$ Pa. The time step is chosen at $\Delta t = 90$ s, using the free-surface version of the model.

4.6.4 Results: Scenario 1

The wind stress imposed (see Fig. 4.17a) creates a surface Ekman layer in the ocean. Details of the vertical structure of this Ekman layer are not revolved by the relatively coarse vertical grid spacing of 20 m chosen. Nevertheless, the model captures the resultant net horizontal movement, the so-called *Ekman-layer transport*. In the Northern Hemisphere, Ekman-layer transport in the surface ocean is directed 90° to the right with respect to the wind direction. Its magnitude varies with the magnitude of the wind stress. Consequently, the wind-stress forcing of Scenario 1 creates a lateral divergence of surface Ekman-layer transports. With ignorance of flows below the surface Ekman layer, this divergence would result in a drop of the sea surface at a rate of:

$$w_{\text{ek}} = \frac{1}{\rho_o f} \frac{\partial \tau_y^{\text{wind}}}{\partial x} \tag{4.24}$$

called *Ekman-pumping velocity*. In this exercise, the Ekman-pumping velocity attains a maximum value in the middle of the model domain of about 6×10^{-3} mm/s or 50 cm per day. The sea surface, however, does not drop at this rate, given that flow divergence in the surface Ekman layer is partially compensated by a convergence of lateral flow in the ocean underneath. Instead, the sea level rapidly approaches a steady state, accompanied by a steady-state geostrophic flow in the surface layer. As a consequence of this, the Ekman-pumping velocity translates now to the vertical displacement speed of the pycnocline caused by lateral flow divergence in the ocean interior. Accordingly, we can anticipate that the shape of the sea level mirrors the horizontal distribution of the Ekman-pumping velocity; that is,

$$\eta \propto w_{\text{ek}} = \frac{1}{\rho_o f} \frac{\partial \tau_y^{\text{wind}}}{\partial x} = -\pi \frac{\tau_o}{\rho_o f L} \sin(\pi x / L) \tag{4.25}$$

which we yield with insertion of Eq. (4.23) into Eq. (4.24). The model prediction reproduces this shape (Fig. 4.19). The sea level attains a depression of >50 cm in the middle of the model domain after 30 days of simulation. The resultant sea-level gradients trigger surface geostrophic flows of maximum speeds of 40 cm/s running into the same direction as the wind (Fig. 4.20). Recall that geostrophic flows run along lines of constant pressure and therefore normal to the model slice. Interestingly, the

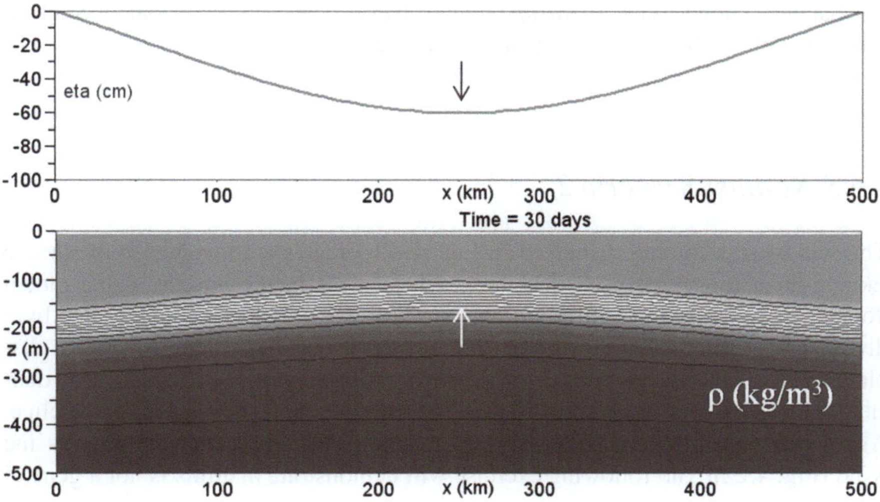

Fig. 4.19 Exercise 19. Scenario 1. Snapshots of sea-level elevation (*top panel*) and density field (*bottom panel*) after 30 days of simulation

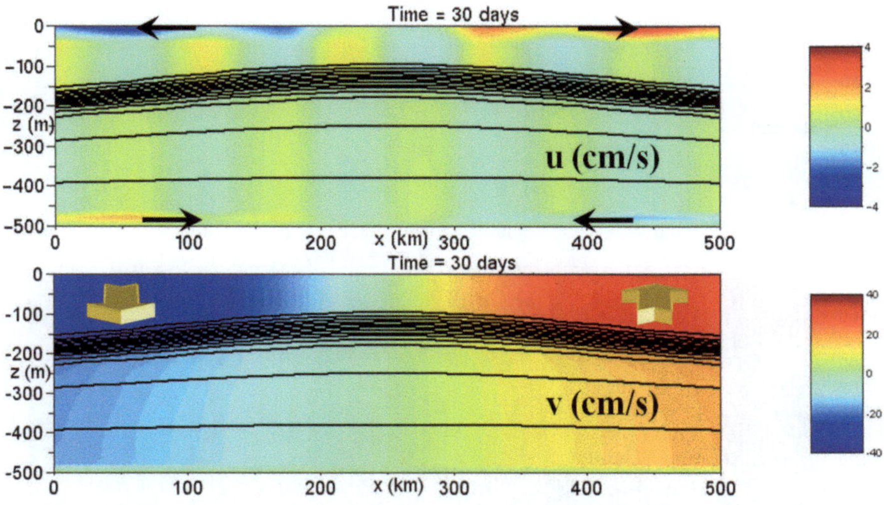

Fig. 4.20 Exercise 19. Scenario 1. Distributions of horizontal velocity components (color shading) after 30 days of simulation. *Lines* represent density contours

wind-stress forcing imposed creates low-amplitude internal waves at the pycnocline. Baroclinic compensation is evident from the weakening of geostrophic flows below the pycnocline.

In the real situation and away from the western boundaries of ocean basins, the divergence of transports in the surface Ekman layer is compensated by a convergence of geostrophic flow owing to meridional variation of the Coriolis parameter; that is, the so-called β effect. The configuration of this exercise does not reproduce this balance, known as the *Sverdrup balance*. Nevertheless, the principal mechanism inherent with the baroclinic compensation process is the same.

4.6.5 Results: Scenario 2

The wind-stress forcing displayed in Fig. 4.17b creates a convergence of surface Ekman-layer transports such that the sea level rises in the middle of the model domain (Fig. 4.21). In equilibrium, this surface convergence is compensated by a divergence of lateral flow in the bottom Ekman layer, inducing a downward displacement of the pycnocline. As in Scenario 1, modulation of the internal density field leads to a weakening of the geostrophic flow below the pycnocline. Again, the geostrophic surface flow runs largely into the same direction as the wind (Fig. 4.22). The following exercise will demonstrate that this is not a general rule.

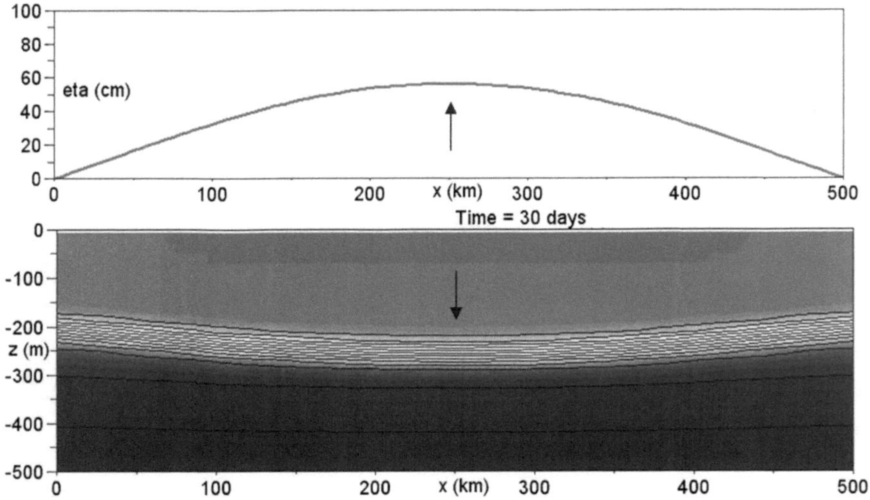

Fig. 4.21 Exercise 19. Same as Fig. 4.19, but for Scenario 2

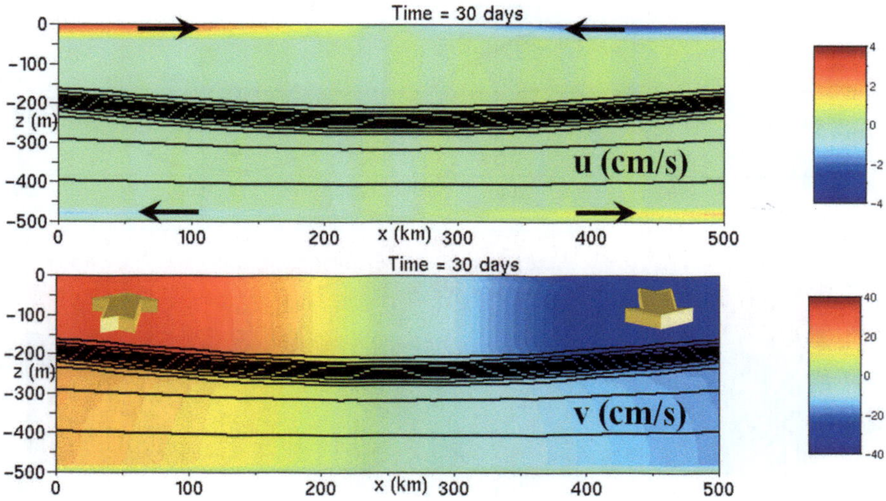

Fig. 4.22 Exercise 19. Same as Fig. 4.20, but for Scenario 2

4.6.6 Results: Scenario 3

Would you believe that winds can produce currents that run against the wind direction? Wouldn't this be a paradox being in gross conflict with expectation? The aim of this exercise is to prove that, on the rotating Earth, the wind can indeed create such flows.

The wind-stress forcing imposed in the scenario (see Fig. 4.17c) creates convergence of surface Ekman-layer transports in the left half of the model domain and divergence in the right half. The distribution of sea-level elevation and, as an amplified mirror image, the distribution of pycnocline depth reflect this (Fig. 4.23). The geostrophic balance for our 2.5d vertical ocean-slice model reads:

$$v = \frac{1}{f\rho_o}\frac{\partial P}{\partial x} \tag{4.26}$$

where P is dynamic pressure. In vicinity of the sea surface, the density-variable part can be neglected and the geostrophic balance for the surface flow reads:

$$v = \frac{g}{f}\frac{\partial \eta}{\partial x} \tag{4.27}$$

where η is sea-level elevation. The latter balance implies that the geostrophic flow in central regions of the model domain, being indirectly created by the wind stress, has to run opposite to the wind direction, since the slope of the sea level is reversed in this region. The speed of this flow even peaks in a region where the wind stress vanishes.

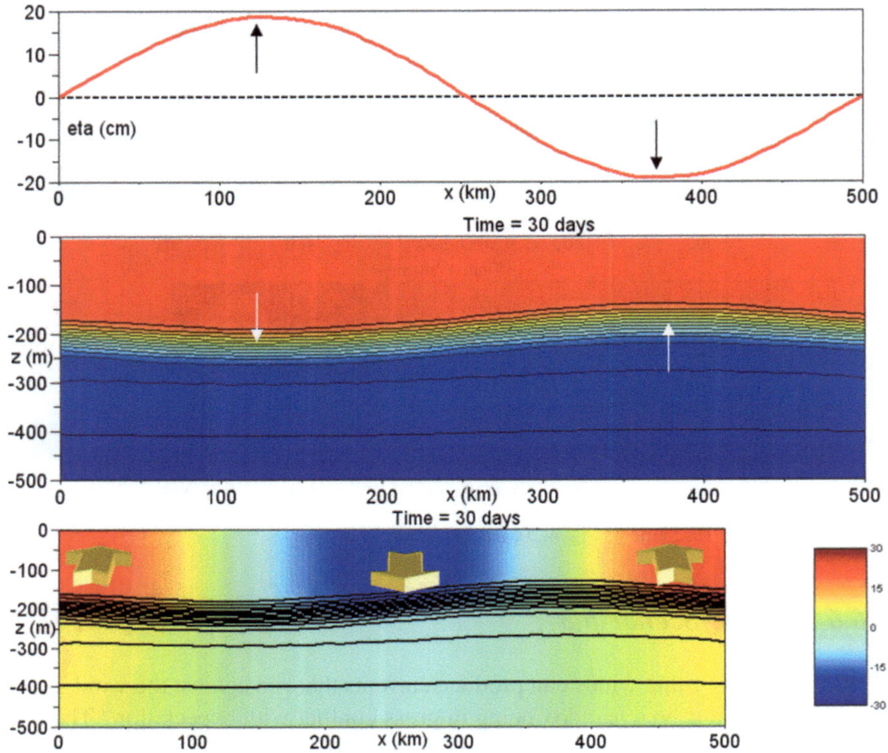

Fig. 4.23 Exercise 19. Scenario 3. Distributions of (*top panel*) sea-level elevation, (*middle panel*) density field, and (*bottom panel*) v-component of velocity after 30 days of simulation. *Black lines* are density contours

This example illustrates the subtle way by which the ocean responds to wind forcing. Equatorial countercurrents are an example of such exotic wind-driven geostrophic flows.

4.6.7 Additional Exercises for the Reader

Repeat Scenario 1 of this exercise with a longer simulation time to explore whether the pycnocline can outcrop at the sea surface, and if so, how this modifies the dynamics. Repeat Scenario 3 of this exercise with the settings of $\tau_o = -0.1\,\text{Pa}$ and $\tau_1 = 0.05\,\text{Pa}$ in Eq. (4.25) and identify regions in which the geostrophic flow runs opposite to the wind direction.

Chapter 5
3D Level Modelling

Abstract This chapter introduces the reader to fully three-dimensional hydrodynamic level models. Exercises address isolated oceanic eddies, eddy formation in a strait, density-driven exchange flows through a strait, coastal upwelling in three dimensions, the thermohaline deep circulation in the ocean, and, as a highlight, equatorially trapped Kelvin waves accompanying El-Niño events.

5.1 The Basic Equations

5.1.1 The Basics

This section outlines the full three-dimensional equations governing the dynamics of an incompressible fluid with a free surface. This comprises the momentum equations, the continuity equation, and a density-conservation equation, plus boundary conditions.

5.1.2 Conservation of Momentum

As in the vertical-ocean slice model, dynamic pressure is separated into two parts:

$$P = p + q \tag{5.1}$$

where (lower-case) p refers to the hydrostatic pressure field with reference to an undisturbed (horizontal) sea level, and q includes pressure effects imposed by a tilted sea surface and by nonhydrostatic pressure contributions. Again, the first pressure part p is diagnosed from the hydrostatic balance:

$$\frac{\partial p}{\partial z} = -(\rho - \rho_o)\, g \tag{5.2}$$

J. Kämpf, *Advanced Ocean Modelling*, DOI 10.1007/978-3-642-10610-1_5,

where z is the vertical coordinate, ρ is true density, ρ_o is surface density, g is acceleration due to gravity. The second pressure contribution q appears implicitly in the momentum equations and cannot be derived in an explicit manner. The momentum equations can be written as:

$$\begin{aligned}\frac{\partial u}{\partial t}+\mathrm{Adv}(u)-fv&=-\frac{1}{\rho_o}\frac{\partial(p+q)}{\partial x}+\mathrm{Diff}(u)\\ \frac{\partial v}{\partial t}+\mathrm{Adv}(v)+fu&=-\frac{1}{\rho_o}\frac{\partial(p+q)}{\partial y}+\mathrm{Diff}(v)\\ \frac{\partial w}{\partial t}+\mathrm{Adv}(w)&=-\frac{1}{\rho_o}\frac{\partial q}{\partial z}+\mathrm{Diff}(w)\end{aligned} \tag{5.3}$$

where (u, v, w) are components of velocity in a Cartesian coordinate system with coordinates (x, y, z), t is time, f is the Coriolis parameter, and ρ_o is reference density. Notice that the pressure split makes the buoyancy force disappear from the vertical momentum equation. In the real situation, the Coriolis force includes a contribution inducing a momentum transfer between the u and w components. This component is often negligibly small (see Cushman-Roisin, 1994). It is therefore not included in the above equations.

The operator Adv() in Eq. (5.3) denotes the advection terms and is given by:

$$\mathrm{Adv}(\psi)=u\frac{\partial\psi}{\partial x}+v\frac{\partial\psi}{\partial y}+w\frac{\partial\psi}{\partial z}$$

where ψ is the property subject to advection. Diffusion of any of the three velocity components is given by:

$$\mathrm{Diff}(\psi)=\frac{\partial}{\partial x}\left(A_h\frac{\partial\psi}{\partial x}\right)+\frac{\partial}{\partial y}\left(A_h\frac{\partial\psi}{\partial y}\right)+\frac{\partial}{\partial z}\left(A_z\frac{\partial\psi}{\partial z}\right)$$

where A_h and A_z are horizontal and vertical eddy viscosities, parameterising effects of turbulence. The separation into horizontal and vertical components is justified given that horizontal mixing is rather triggered by larger-size geostrophic eddies, whereas turbulence via vertical shear of lateral currents drives mixing in the vertical.

5.1.3 Conservation of Volume

Conservation of volume is expressed by the continuity equation for an incompressible fluid that can be written as:

$$\frac{\partial u}{\partial x}+\frac{\partial v}{\partial y}+\frac{\partial w}{\partial z}=0 \tag{5.4}$$

Vertical integration of the continuity equation gives a prognostic equation for surface dynamic pressure, yielding:

$$\frac{\partial q_s}{\partial t} = -\rho_o g \left(\frac{\partial (h \langle u \rangle)}{\partial x} + \frac{\partial (h \langle v \rangle)}{\partial y} \right) \tag{5.5}$$

where $q_s = \rho_o g \eta$ with ρ_o being surface density and η being sea-surface elevation, h is total local water depth, and $\langle . \rangle$ represents a vertical average.

5.1.4 Evolution of the Density Field

Seawater density depends on temperature, salinity and pressure. For simplicity, we assume a linear dependence of seawater on temperature and salinity, ignore any pressure effects, and assume that eddy diffusivities for heat and salt are the same. To this end, the evolution of the density field can be described by a density-conservation equation, given by:

$$\frac{\partial \rho}{\partial t} + \text{Adv}(\rho) = \text{Diff}(\rho) \tag{5.6}$$

where the diffusion operator is given by:

$$\text{Diff}(\rho) = \frac{\partial}{\partial x}\left(K_h \frac{\partial \rho}{\partial x} \right) + \frac{\partial}{\partial y}\left(K_h \frac{\partial \rho}{\partial y} \right) + \frac{\partial}{\partial z}\left(K_z \frac{\partial \rho}{\partial z} \right)$$

where K_h and K_z, respectively, are horizontal and vertical eddy diffusivities.

5.2 Numerical Treatment

5.2.1 The 3d Arakawa C-grid

The index triplet (i, j, k) is used as a pointer to certain grid cells of the Arakawa C-grid (Fig. 5.1), where Δx is the grid spacing in the x-direction, Δy is the grid spacing in the y-direction, and Δz is the grid spacing in the vertical direction. Note that the i index runs opposite to the z-coordinate. For convenience, we locate the Cartesian coordinate system such that the x-axis points to the east, the y-axis to the north, and the z-axis upward. Grid points of scalars (pressure, density, Eulerian concentration, eddy viscosity, etc.) are centred between velocity grid points. Within each grid cell, the u-grid point is located to the east, the v-grid point to the north, and the w-grid point above with respect to the scalar grid point.

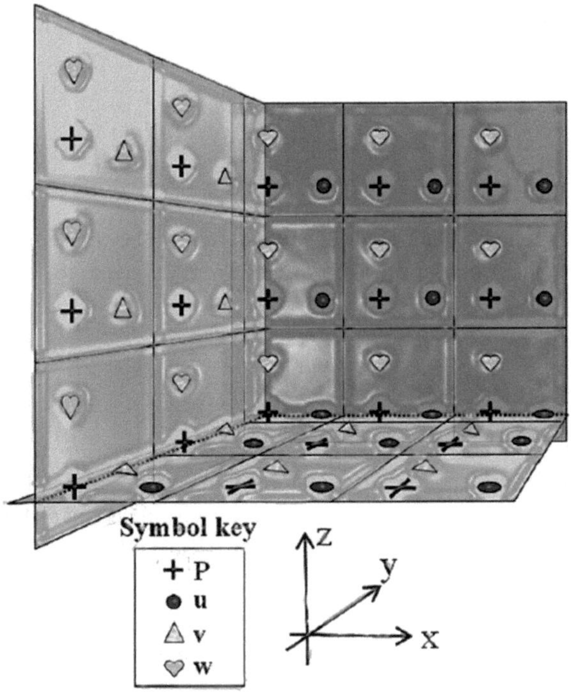

Fig. 5.1 The three-dimensional Arakawa C-grid

5.2.2 Treatment of the Advection Terms

Using the product rule of differentiation, the advection terms for a property B can be written as:

$$u\frac{\partial B}{\partial x}+v\frac{\partial B}{\partial y}+w\frac{\partial B}{\partial z}=\frac{\partial(uB)}{\partial x}+\frac{\partial(vB)}{\partial y}+\frac{\partial(wB)}{\partial z}-B\left(\frac{\partial u}{\partial x}+\frac{\partial v}{\partial y}+\frac{\partial w}{\partial z}\right) \quad (5.7)$$

The last term involves the continuity equation (Eq. 5.4) and therefore should vanish in theory. Nevertheless, this term is retained in the numerical advection scheme for elimination of small round-off errors that otherwise could accumulate during a simulation. A control-volume approach is used to calculate advection of scalars and the nonlinear terms in the momentum equations (Fig. 5.2). With this approach, fluxes of a property into this control volume are calculated from fluxes through its east/west, north/south and top/bottom faces, respectively. Again, individual fluxes are computed by means of a Total-Variation-Diminishing (TVD) advection scheme using the Superbee limiter (see Sect. 3.6). Implementation for the newly introduced y-direction should be straight forward.

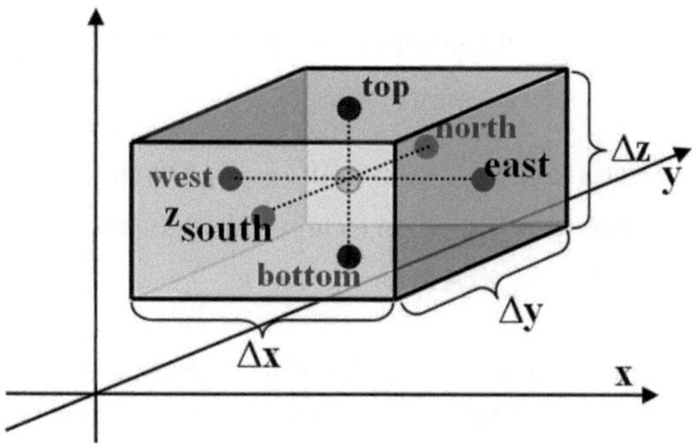

Fig. 5.2 The three-dimensional control volume

5.2.3 The Nonhydrostatic Solver of the Momentum Equations

As in the vertical ocean-slice model, the S.O.R. scheme is used to derive the pressure field q. To this end, q is split up into explicit and implicit components; that is,

$$q \Rightarrow q^n + \Delta q^{n+1}$$

For the fully three-dimensional momentum equations, the Poisson equation for Δq reads:

$$a_e \Delta q^{n+1}_{i,j,k+1} + a_w \Delta q^{n+1}_{i,j,k-1} + a_n \Delta q^{n+1}_{i,j+1,k} + a_s \Delta q^{n+1}_{i,j-1,k} + \\ + a_t \Delta q^{n+1}_{i-1,j,k} + a_b \Delta q^{n+1}_{i+1,j,k} - a_o \Delta q^{n+1}_{i,j,k} = q^*_{i,j,k} \quad (5.8)$$

The coefficients in this equation for uniform grid spacings are given by:

$$\begin{aligned} a_e &= a_w = \Delta z / \Delta x \\ a_n &= a_s = \Delta z \Delta x / (\Delta y)^2 \\ a_t &= a_b = \Delta x / \Delta z \\ a_o &= a_e + a_w + a_n + a_s + a_t + a_b \end{aligned}$$

The right-hand side of the Poisson equation (Eq. 5.8) represents the divergence of first-guess values of velocity and is given by:

$$q^*_{i,j,k} = \frac{\rho_o}{\Delta t} \Big[(u^*_{i,j,k} - u^*_{i,j,k-1}) \Delta z + \\ + (v^*_{i,j,k} - v^*_{i,j-1,k}) \frac{\Delta x \Delta z}{\Delta y} + (w^*_{i,j,k} - w^*_{i+1,j,k}) \Delta x \Big] \quad (5.9)$$

First-guess values of velocity components in the latter equation are calculated from:

$$
\begin{aligned}
u^*_{i,j,k} &= \cos(\alpha)u^n_{i,j,k} + \sin(\alpha)v^n_u - \Delta t\mathrm{Adv}(u) + \Delta t\ F^n_u \\
v^*_{i,j,k} &= \cos(\alpha)v^n_{i,j,k} - \sin(\alpha)u^n_v - \Delta t\mathrm{Adv}(v) + \Delta t\ F^n_v \\
w^*_{i,j,k} &= w^n_{i,j,k} - \Delta t\mathrm{Adv}(w) + \Delta t\ F^n_w
\end{aligned}
\tag{5.10}
$$

where $\alpha = \Delta t f$, u_v are u-values interpolated to v-grid points and v_u are v-values interpolated to u-grid points. Finite-difference versions of the nonlinear terms are represented by Adv(u), Adv(v), and Adv(w). The Coriolis force is implemented using the local-rotation approach (see Sect. 4.1). The remaining terms in Eq. (5.10) are given by:

$$
\begin{aligned}
F^n_u &= -\frac{1}{\rho_o \Delta x}(p^n_{i,j,k+1} - p^n_{i,j,k} + q^n_{i,j,k+1} - q^n_{i,j,k}) + \mathrm{Diff}(u) \\
F^n_v &= -\frac{1}{\rho_o \Delta y}(p^n_{i,j+1,k} - p^n_{i,j,k} + q^n_{i,j+1,k} - q^n_{i,j,k}) + \mathrm{Diff}(v) \\
F^n_w &= -\frac{1}{\rho_o \Delta z}(q^n_{i-1,j,k} - q^n_{i,j,k}) + \mathrm{Diff}(w)
\end{aligned}
\tag{5.11}
$$

where Diff(u), Diff(v), and Diff(w) denote finite-difference versions of momentum diffusion terms. Once the S.O.R. iteration has converged to a user-specified pressure accuracy, values of velocity components at the next time level ($n + 1$) are given by:

$$
\begin{aligned}
u^{n+1}_{i,j,k} &= u^*_{i,j,k} - \frac{\Delta t}{\rho_o \Delta x}(\Delta q^r_{i,j,k+1} - \Delta q^r_{i,j,k}) \\
v^{n+1}_{i,j,k} &= v^*_{i,j,k} - \frac{\Delta t}{\rho_o \Delta y}(\Delta q^r_{i,j+1,k} - \Delta q^r_{i,j,k}) \\
w^{n+1}_{i,j,k} &= w^*_{i,j,k} - \frac{\Delta t}{\rho_o \Delta z}(\Delta q^r_{i-1,j,k} - \Delta q^r_{i,j,k})
\end{aligned}
\tag{5.12}
$$

where the index r refers to the result of the S.O.R. iteration. Analog to treatment in the vertical ocean-slice model, the three-dimensional S.O.R. iteration includes updates of the surface value of Δq via calculation of the lateral divergence of depth-integrated values of u and v.

5.2.4 Stability Criteria

The CFL stability criterion associated with advection of a property is given by:

$$
\Delta t \le \min\left(\frac{\Delta x}{u}, \frac{\Delta y}{v}, \frac{\Delta z}{w}\right) \tag{5.13}
$$

It can be shown (Kowalik and Murty, 1999) that the CLF stability associated with the propagation of surface gravity waves is given by:

$$\Delta t \leq \frac{\min(\Delta x, \Delta y)}{\sqrt{2gh_{\max}}} \tag{5.14}$$

where $h_{\max}$ is the maximum water depth of the model domain. Notice that the factor of two appears here in the denominator. The latter condition is relevant for applications considering a free sea surface. Longer numerical time steps may work with choice of the rigid-lid approximation (see Sect. 3.7). In principle, the three-dimensional free-surface hydrodynamic level model, described here, can be employed to simulate any of the previous exercises in three-dimensional space. Nevertheless, to avoid super-long simulation times, the reader should keep the number of grid points as small as possible. Also data output should be restricted to a few selected horizontal and vertical transects of variables to avoid data-storage problems.

5.3 Exercise 20: Geostrophic Adjustment in 3D

5.3.1 Aim

The aim of this exercise is to test and validate the three-dimensional free-surface hydrodynamic level model by means of the geostrophic adjustment problem using a configuration similar to that of Exercise 15.

5.3.2 Task Description

The model domain is 5 × 5 km in lateral extent and 500 m in depth (Fig. 5.3). Lateral grid spacings are set to $\Delta x = \Delta y = 2$ km. The vertical grid spacing is set to $\Delta z = 20$ m. This gives a total of 25×25×25 =15,625 grid points, which exceeds by far the number of grid points used in previous model simulations. Zero-gradient conditions are used for all variables at lateral boundaries.

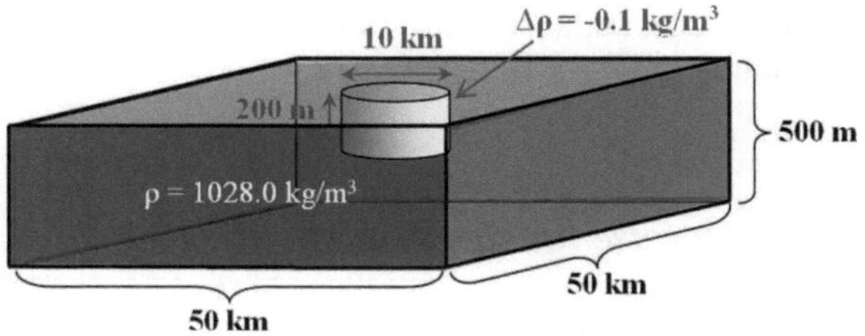

Fig. 5.3 Initial configuration for Exercise 20

The model is forced by initial prescription of a cylindrical patch of surface water of 10 km in diameter and a thickness 200 m. This patch is initially 0.1 kg/m^3 lighter compared with ambient water of a density of ρ_o = 1,028 kg/m^3. The associated density anomaly is linearly adjusted from zero to its final value during the first 2 hrs of simulation.

The Coriolis parameter is set to $f = 1 \times 10^{-4}\,\mathrm{s}^{-1}$. We expect that geostrophic adjustment creates frontal currents running around the rim of the low-density patch. We also expect that the width of this current is of the order of the internal Rossby radius of deformation, given by Eq. (4.12), which is approximately 3.4 km in this exercise. Although the lateral grid spacing of 2 km does not adequately resolve this length scale, the results will show that the model is able to capture key dynamical aspects of the geostrophic adjustment process.

Wind forcing is not applied in this exercise. Horizontal eddy viscosity and eddy diffusivity are set to uniform values of $A_h = K_h = 1\,\mathrm{m}^2/\mathrm{s}$. Kochergin's turbulence closure scheme, Eq. (4.22), is employed for calculation of variable vertical eddy viscosity and eddy diffusivity. The bottom-friction parameter in the assumed quadratic bottom-friction law is set to $r = 0.001$.

The total simulation time is 60 hrs (2.5 days) with data outputs at hourly interval. Data outputs are those of surface distributions of density, horizontal velocity components and sea-level elevation, and vertical transects of density and horizontal velocity components across the centre of the model domain at y = 25 km. The time step is set to Δt = 5 s, using the free-surface version of the model. Pressure accuracy of the S.O.R. iteration is set to ϵ = 0.01 Pa.

5.3.3 Results

The geostrophic adjustment process creates a high-pressure centre associated with a sea-level elevation of 1 cm (not shown). This drives an anticyclonic geostrophic surface eddy of approximately 20 cm/s in speed (Fig. 5.4) superimposed on which are inertial oscillations. Instead of continued lateral spreading, the Coriolis force operates to maintain the low-density surface patch as a circular feature that only slowly dissipates owing to both lateral density diffusion and frictional effects. As anticipated, a cyclonic eddy establishes in the bottom layer owing to water-column stretching (Fig. 5.5). In summary, the three-dimensional model appears to capture key dynamical aspects of the geostrophic adjustment process, even with a relatively coarse spatial resolution. Note the striking similarity of the 3d findings with those of the 2.5d application in Exercise 16 (see Sect. 4.2).

5.3.4 Additional Exercise for the Reader

Place a cylindrical patch of denser water at the bottom of the model domain and explore the geostrophic adjustment process that follows for this configuration. Vary

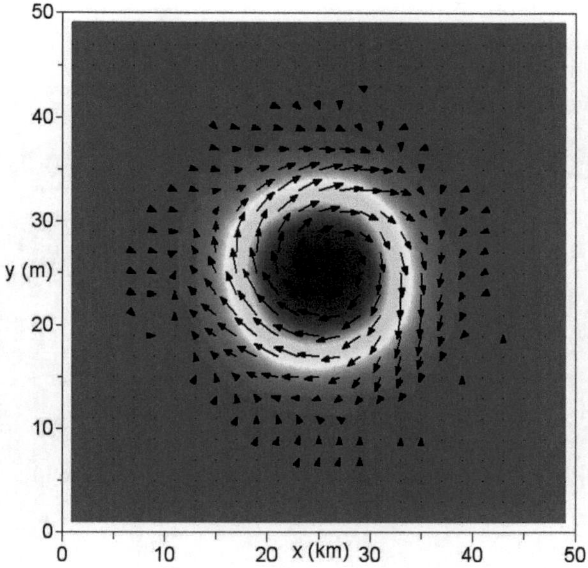

Fig. 5.4 Exercise 20. Surface distribution of density (*shading*) and horizontal velocity (*arrows*) after 2 days of simulation. Velocity grid points of speeds <1 cm/s are omitted

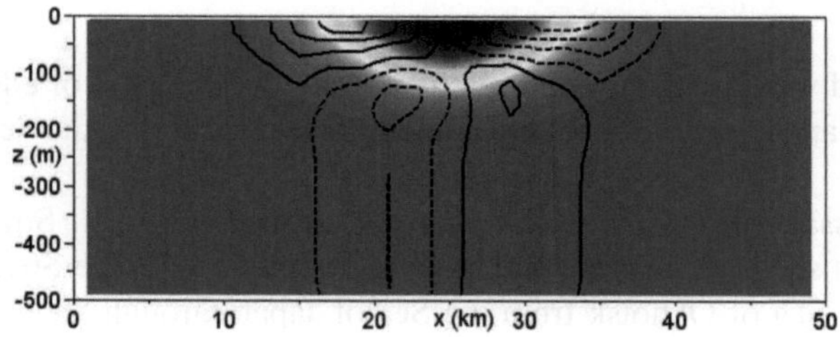

Fig. 5.5 Exercise 20. Vertical transect at $y = 25$ km showing density (*shading*) and the v-velocity component (contours at 4 cm/s interval) after 2 days of simulation. *Solid* (*broken*) *lines* refer to positive (negative) values of v

the value of the bottom-friction parameter over one order of magnitude to explore the role that bottom friction and the bottom-Ekman layer play in the dynamics evolving.

5.4 Exercise 21: Eddy Formation in a Strait

5.4.1 Background

Eddy shedding by frontal flows by the barotropic and/or baroclinic instability mechanisms is a common feature in the ocean. For physical details of these instability mechanisms the reader is referred to other books (e.g., Cushman-Roisin, 1994). In

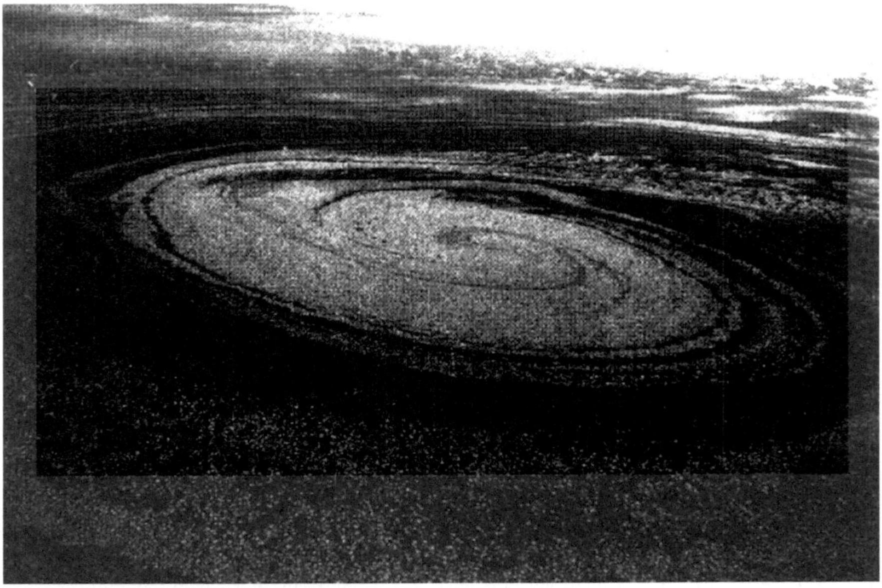

Fig. 5.6 An ocean eddy of about 20 km in radius. The spirals are ice bands. The photo was taken from an airplane flying over the Sea of Okhotsk. From Wakasutchi and Ohshima (1990)

regions of cold winter climates and sea ice formation, the trace of eddies can produce spectacular and beautiful patterns in the sea ice, such as in the Sea of Okhotsk (Fig. 5.6).

Narrow passages connecting ocean basins are called *straits*. Straits are often combined with shallower water depth associated with a *sill*. The Soya Warm Current enters the Sea of Okhotsk from the Sea of Japan through the Soja Strait and continues along the coast of Hokkaido as a coastal boundary current. Instability of this coastal current produces spiralling ice patterns which are akin to cloud patterns produced by hurricanes.

5.4.2 Aim

The aim of this exercise is to simulate eddy shedding of the Soya Warm Current in the Sea of Okhotsk with the three-dimensional hydrodynamic model developed in this book. To this end, an experimental design similar to that used by Ohshima and Wakatsuchi (1990) in their numerical study will be employed.

5.4.3 Task Description

This exercise requires variable bathymetry resembling that of the Sea of Okhotsk. One option would be to download realistic gridded bathymetric data from the Internet. The General Bathymetric Chart of the Oceans (GEBCO) found at http://www.gebco.net would be a useful data source. Instead of this, the author wishes to demonstrate the reader an alternative way to create an idealised bottom topography using bathymetric charts as visual template. Section 5.4.4 details the method creating the bathymetry shown in Fig. 5.7.

The model domain is 360 km long and 180 km across resolved by a lateral isotropic grid spacing of 3.6 km. The objective here is to be able to just resolve the eddy scale of 30 km with a sufficient number of grid points while the model domain is large enough to capture several eddies. The islands Hokkaido and Sakhalin appear as rectangular blocks, which is sufficient for the purpose of the study. The model's Soya Strait has a width of about 60 km and a maximum depth of 60 m. Water depth in the Sea of Okhotsk is limited to 200 m to allow for relatively large numerical time steps. The vertical grid spacing is set to 20 m. The Coriolis parameter is set to $1 \times 10^{-4}\,\mathrm{s}^{-1}$, corresponding to an inertial period of 17.45 hrs.

For simplicity, density is assumed uniform and the existence of sea ice is ignored. The model is forced via prescription of an inflow from the Japan Sea with an arbitrarily chosen speed of 20 cm/s. This speed is applied to velocity components directed normal to the open boundary. Parallel velocity components are kept at zero value. To avoid initial disturbances, the inflow speed is gradually adjusted to its final value over the first 2 days of simulation. Zero-gradient conditions are used for other variables at inflow boundaries. At the downstream open boundary, zero-gradient conditions are used for all variables.

Lateral momentum diffusion with uniform eddy diffusivity of $A_h = 5\,\mathrm{m^2/s}$ is used in conjunction with *no-slip boundary conditions* for flow parallel to coastlines. This condition is implemented via appropriate settings of the velocity value on the first inland grid point (see Fig. 5.8 for an example). To avoid problems in the ocean interior, this condition is only used in grid cells adjacent to coastlines. No-slip conditions are used adjacent to steps in bathymetry. The Kochergin scheme is employed for parametrisation of vertical turbulence using the same parameter settings as in Exercise 20. For simplicity, bottom friction is disabled.

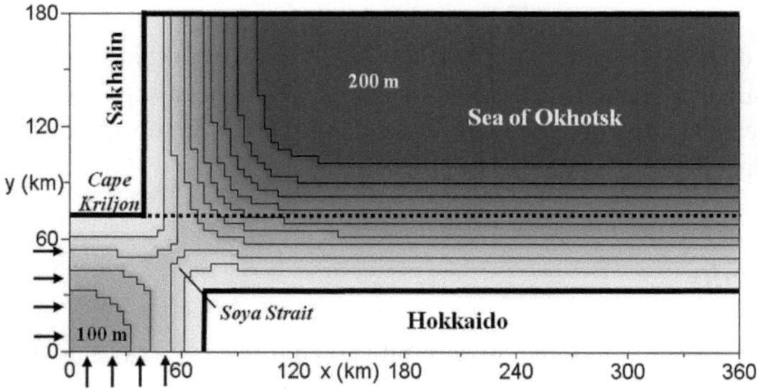

Fig. 5.7 Model geometry for Exercise 21. *Thick solid lines* denote closed boundaries. *Arrows* indicate the inflow boundaries. The *dashed line* indicates a boundary used for the initial prescription of Eulerian tracer concentration

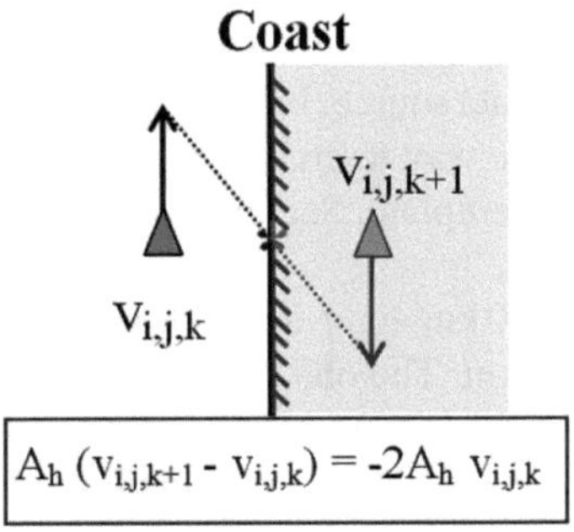

Fig. 5.8 Example of the implementation of no-slip conditions for flow parallel to coastlines

A field of Eulerian tracer concentration is added for visualisation of the fluid dynamics. To this end, concentrations of unity are allocated initially on one side of the dashed line in Fig. 5.7, whereas zero values are assigned on the other side. Patterns evolving in this tracer field are indicative of possible sea ice patterns forming in this region. The assumption here is that sea ice operates as a passive tracer which is only valid for young stages of the ice-formation process; that is, before a solid ice sheet has been formed. The total simulation time is 10 days with a numerical time step of $\Delta t = 45\,\text{s}$. The pressure accuracy for the S.O.R. simulation is set to $\epsilon = 0.01\,\text{Pa}$.

5.4.4 Creation of Variable Bathymetry

Variable bathymetry can be created via prescription of initial coastlines and block-type regions of certain water depths and the use of a diffusion equation for subsequent smoothing. The diffusion equation is given by:

$$\frac{\partial h}{\partial t} = \kappa \left(\frac{\partial^2 h}{\partial x^2} + \frac{\partial^2 h}{\partial y^2} \right) \tag{5.15}$$

where the diffusion coefficient κ and the duration of smoothing are adjusted such that the result is acceptable. Coastlines and land should not disappear during the process. This can be implemented in the code via the choice of zero-gradient conditions at the borders between dry and wet grid cells.

5.4.5 Results

The inflow passes through the Soya Strait and continues as a coastal boundary current along the coast of Hokkaido (Fig. 5.9). The current approaches a maximum speed of 1.0–1.3 m/s at a distance of 20–25 km from the coast, which agrees with observational evidence. The barotropic flow (baroclinic effects are eliminated via choice of uniform density) becomes dynamically unstable and forms meanders in

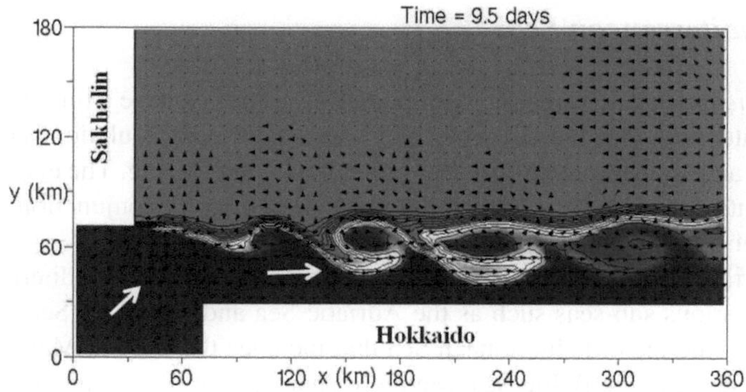

Fig. 5.9 Exercise 21. Snapshot of the distributions of Eulerian tracer (*shading* and *contours*) and the flow field (arrows). Flow vectors are interpolated over 2 × 2 grid cells

interaction with the irregular coastline of Sakhalin and bathymetric variations. Some meanders develop into isolated eddies of about 30 km in diameter. All eddies formed are cyclonic features of anti-clockwise rotation. This sign of relative vorticity and the spatial scale of mesoscale eddies are in agreement with the ice-spiral pattern shown in Fig. 5.7. Eddies are not stationary features. Instead of this, eddies are rapidly carried away by the simulated Soya Warm Current.

5.4.6 *Bathymetry Creation*

The folder "Miscellaneous/Bathymetry for Exercise 21" on the book's ftp site contains the FORTRAN bathymetry creator, named "BathCreator.f95", needed for this exercise. The result is written to a file named "topo.dat" that is requires as input file for the FORTRAN simulation code. Included is also the SciLab script, called "Bath.sce" that creates Fig. 5.7.

5.4.7 *Additional Exercises for the Reader*

Repeat this exercise with the same coastline configuration but uniform water depth of 100 m. Does this configuration also support eddy formation? Include bottom friction and explore variations in the dynamic response.

5.5 Exercise 22: Exchange Flow Through a Strait

5.5.1 *Aim*

The aim of this exercise is to study density-driven exchange flows through a strait that connects ocean basins containing water of different densities.

5.5.2 Mediterranean Seas

Mediterranean seas are semi-enclosed bodies of water that have a limited exchange of deep water with the ambient ocean and where the water circulation is dominated by salinity and temperature differences rather than wind forcing. The exchange with the adjacent ocean is constrained by a narrow strait often in conjunction with shallower water depth associated with a *sill*.

Mediterranean seas of the Atlantic Ocean are the Eurafrican Mediterranean Sea including various sub-seas such as the Adriatic Sea and the Black Sea, the Arctic Ocean, the American Mediterranean Sea that includes the Gulf of Mexico and the Carrabian Sea, and the Baltic Sea. Mediterranean seas of the Indian Ocean are the Persian Gulf and the Red Sea, and the Australasian Mediterranean Sea, including the Banda, Sulu, Sulawesi and Java Seas, being connected with the Pacific Ocean. See (Tomczak and Godfrey, 2003) for more details.

Mediterranean seas share characteristics of estuaries, but the latter are generally much shallower and narrower including gulfs and seaward portions of rivers. Some scientists including the author prefer to classify the Baltic Sea and the Persian Gulf as estuaries.

Mediterranean seas receiving more freshwater via continental runoff and direct precipitation than they lose by evaporation are called *dilution basins* and are found in humid climatic zones. The surplus of freshwater in these systems creates an outflow of low-salinity water in the surface layer and an inflow of oceanic water in the lower layer (Fig. 5.10a). On the other hand, *concentration basins* experience a net freshwater loss owing to excessive evaporation and are situated in arid climatic zones. The loss of freshwater from these systems leads to an outflow of saline sub-surface

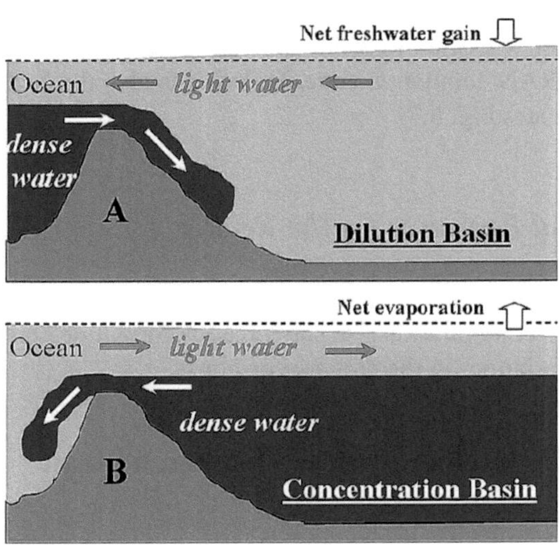

Fig. 5.10 The circulation of mediterranean seas for (*top panel*) dilution basins and (*bottom panel*) concentration basins

water and to an inflow of surface water from the ambient ocean (Fig. 5.10b). The Eurafrican Mediterranean Sea, the Red Sea and the Persian Gulf are examples of concentration basins. The Baltic Sea and the Arctic Ocean are examples of dilution basins.

The Coriolis force comes into play in this exchange circulation when the width of the entrance to a mediterranean sea exceeds the internal Rossby radius of deformation. In this situation, geostrophic adjustment triggers a predominantly horizontal exchange circulation. In additional to this, geostrophic frontal flows often lead to dynamic instabilities and shedding of eddies, which makes the exchange circulation complex and difficult to predict. On the other hand, relatively narrow entrances constrain the circulation to a purely vertical overturning circulation.

5.5.3 Task Description

The model domain is 200 km in length and 100 km across, resolved by lateral grid spacings of $\Delta x = \Delta y = 4$ km (Fig. 5.11). Included are two separate "deeper" ocean basins, 100 m in depth, being connected by a strait of 50 m in depth. The vertical grid spacing is set to $\Delta z = 10$ m. The strait is 40 km in length and 20 km in width. Initially, water density is uniform at 1,028 kg/m^3 and there are no currents. The exchange circulation is created by gradually increasing the density of the entire water column in the left half of the model domain by $\Delta\rho = 2$ kg/m^3 over the first 2 days of the simulation. This is done via the flux condition:

$$\rho_{i,j,k} = \rho_{i,j,k} + \frac{\Delta t}{T}\Delta\rho$$

where T is taken as 2 days (expressed in seconds), and Δt is the numerical time step. This type of forcing allows for density anomalies to evolve during the

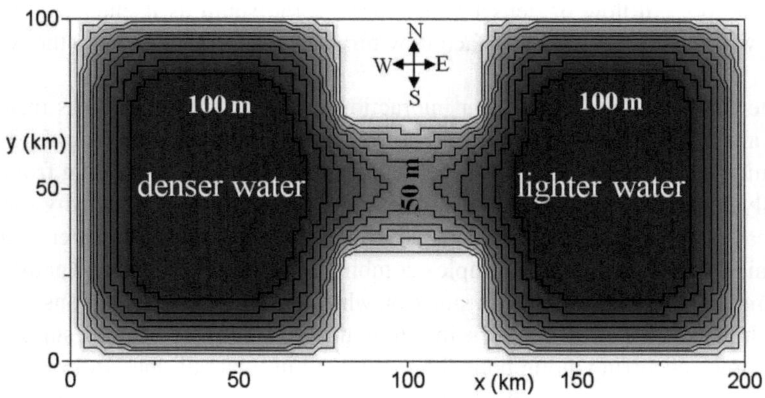

Fig. 5.11 Model configuration for Exercise 22

adjustment period. The Coriolis parameter is set to $f = 1 \times 10^{-4}\,\mathrm{s}^{-1}$ (Northern Hemisphere). This configuration gives an internal deformation radius of approximately 7 km, implying that geostrophic adjustment can take place within length scales imposed by the strait's geometry. Coriolis effects can certainly not be ignored here. The lateral grid spacing only marginally resolves dynamic processes on length scales on the internal deformation radius and some bias in the dynamics is to be expected.

Lateral eddy viscosity and eddy viscosity are given a constant value of $A_h = K_h = 1\,\mathrm{m}^2/\mathrm{s}$. The no-slip condition is used along coastlines. Vertical eddy viscosity and eddy diffusivity, assumed equal, are diagnosed from Kochergin's turbulence closure scheme. The bottom-friction parameter is set to $r = 0.001$. The total simulation time is 20 days with data outputs at every 4 hrs. Data outputs consist of surface and bottom distributions of density and lateral flow fields together with a vertical transect of these variables across the middle of the strait at $x = 100$ km. The free-surface version of the model is used with a time step of $\Delta t = 10$ s, which satisfies the CFL criterion for fast-propagating surface gravity waves (Eq. 5.14). The pressure accuracy for the S.O.R. simulation is set to $\epsilon = 0.01$ Pa.

5.5.4 Results

The density forcing creates a lateral density contrast across the strait. This initiates a bottom-arrested density-driven current moving denser bottom water from the western basin into the eastern basin (Fig. 5.12). Volume transports associated with this current lower the sea level in the western basin and lift the sea level in the eastern basin. The resultant sea-level gradient, in turn, triggers a westward surface return flow through the strait.

The Coriolis force becomes a dominant force within a timescale of a few days. Geostrophic adjustment along density fronts follows. As a result of this adjustment, the bottom flow of denser water turns to the south as it enters the eastern basin, whereas the returning surface flow turns to the north as it enters the western basin.

Later in this process, nonlinear interaction produces transient eddies inside the strait, and a stationary barotropic eddy forms in the outflow region (Fig. 5.13). The dynamics and density structure in the strait are highly transient varying from horizontally aligned density surfaces to situations of vertically aligned density surfaces. The reason behind this intermittency are transient pulses of dense water outflows, dynamical instabilities, and a complex combination of gravitational and geostrophic adjustments. Figure 5.14 shows a situation which agrees with expectations. At other times, however, the outflow can be intermittently concentrated near the surface with return flows occurring along both sides of the strait. Overall, the outflow of dense water gives rise of an anti-clockwise circulation pattern in the eastern basin. In reality, dense outflows can travel vast distances along the continental slope such

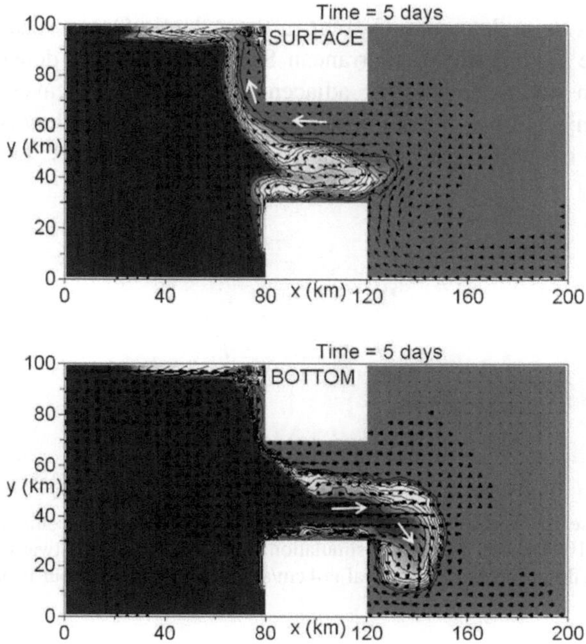

Fig. 5.12 Exercise 22. Surface and bottom horizontal distributions of density (*shading* and *contours*) together with horizontal flow fields (*arrows*) after 5 days of simulation. Velocity grid points of speeds < 1 cm/s are omitted

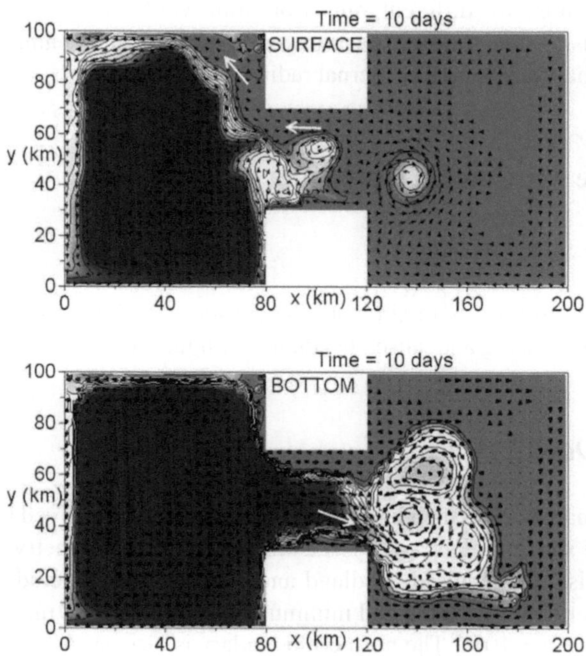

Fig. 5.13 Exercise 22. Same as Fig. 5.12, but after 10 days of simulation

as the *Deep Western Boundary Current* in the Atlantic Ocean. Outflow of dense water from the Eurafrican Mediterranean Sea finds an equal density horizon at around 1000-m water depth in the adjacent ocean where it leaves the sea floor and becomes injected into the ambient water column. This outflow is also popular for its creation of submerged mesoscale eddies called "meddies" by the scientific community.

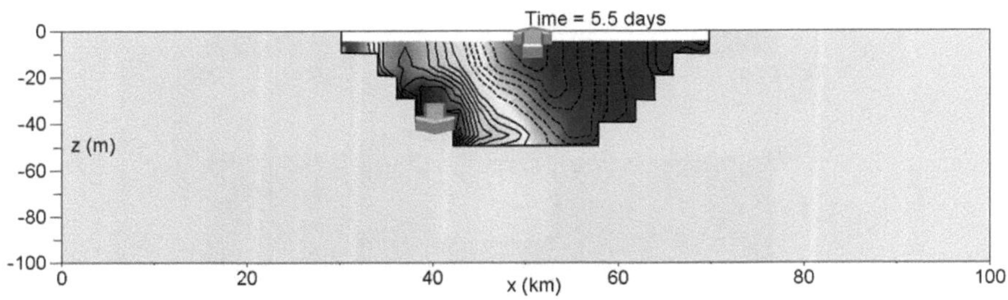

Fig. 5.14 Exercise 22. Vertical transect of density (*shading*) and the u-component of velocity (*contours*) at $x = 100$ km after 5.5 days of simulation. *Solid lines* refer to eastward flow and broken lines to westward flow. The contour interval is 4 cm/s. The zero-value contour is omitted

5.5.5 Additional Exercise for the Reader

Repeat this exercise for different values of strait widths and/or density contrasts between the basins. The reader is encouraged to explore the dynamics evolving for a strait width that is close to the internal radius of deformation.

5.6 Exercise 23: Coastal Upwelling in 3D

5.6.1 Aim

The aim of this exercise is to simulate the coastal upwelling circulation in three spatial dimensions using a idealised headland configuration.

5.6.2 Task Description

The model domain is 200 km in length and 100 km in width resolved by an isotropic horizontal grid spacing of $\Delta x = \Delta y = 2$ km. Using the bathymetry creator of the previous exercise, an idealised headland and a small bay is included (Fig. 5.15). Maximum water depth is 100 m and minimum water depth is 20 m, using a vertical grid spacing of $\Delta z = 10$ m. The northern boundary is a coast. All other boundaries are open boundaries using zero-gradient conditions for all variables. An exception

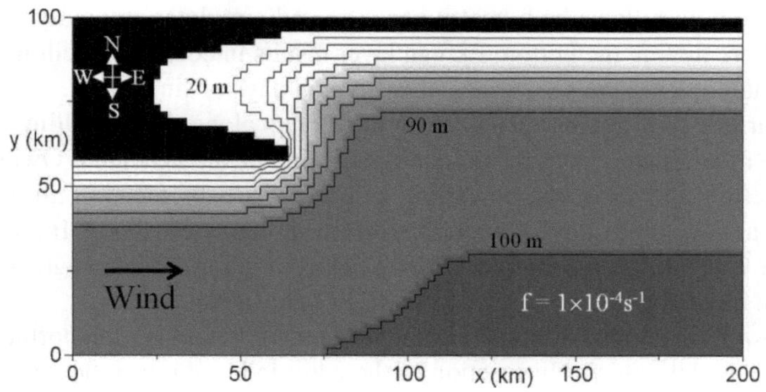

Fig. 5.15 Configuration for Exercise 23

is the sea-level anomaly which is kept at zero value along the southern offshore boundary. No-slip conditions are used for flow parallel to coastlines (see Sect. 5.4).

Initially, the upper 40 m of the water column has a density of ρ_1 = 1,028 kg/m^3. The ocean underneath is by $\Delta\rho$ = 1 kg/m^3 denser. Together with a Coriolis parameter of $f = 1{\times}10^{-4}\,\mathrm{s}^{-1}$ (Northern Hemisphere), this gives a frontal width (take Eq. 4.12) of $R \approx 6.2$ km, which is barely resolved by the lateral grid spacing chosen. Horizontal eddy viscosities and eddy diffusivities are set to uniform values of 1 m^2/s. Kochergin's turbulence closure scheme is used for parametrisation of sub-grid scale vertical mixing. The same settings as in previous exercises are used. The bottom friction coefficient is set to a value of r = 0.001.

The model is forced via prescription of an upwelling favorable eastward wind stress of 0.1 Pa in magnitude. After Eq. (2.6), this corresponds to a wind speed of ~7.2 m/s for an assumed wind-drag coefficient of $C_d = 1.5 \times 10^{-3}$. This wind forcing is gradually blended in over the first two days of simulation. According to Eq. (4.19) and for a wind-stress magnitude of 0.1 Pa, full upwelling is expected to occur after 3 days of simulation. The initial wind-stress adjustment, however, will lead to a slight delay of this expected upwelling response.

Eulerian tracer concentration is used to mark sub-surface shelf water below 40-m depth. The total simulation time is 10 days, being of the order of the typical timescale of weather events, using a numerical time step of 30 secs. The rigid-lid approximation is not employed. The pressure accuracy of the S.O.R. iteration is set to ϵ = 0.01 Pa.

5.6.3 Results

The wind-stress forcing creates an offshore Ekman drift in the surface layer along southward facing stretches of the coast. This offshore drift lowers the coastal sea level and produces a geostrophic flow running into the wind direction. Initially, the geostrophic flow follows bathymetry contours and circulates smoothly around the bay. Onshore flow in the bottom Ekman layer results in coastal upwelling in which deeper water reaches the sea surface after about 5 days of simulation.

The surface flow attains speeds of up to 80 cm/s along the upwelling front that gradually moves away from the southern coastline of the headland. Owing to this departure from the coast, the upwelling jet looses its bathymetry control and produces an anticyclonic eddy of about 25–30 km in diameter (Fig. 5.16, top panel). This scale is 4–5 times the internal Rossby radius of deformation in agreement with baroclinic instability theory (e.g., Cushman-Roisin, 1994).

An upwelling jet also establishes downstream the bay along the northern coastline of the model domain. Interaction of the eddy-induced return flow with this jet leads to formation of a counter-rotating vortex pair (Fig. 5.16, bottom panel). When recalling the direction of Ekman drift in the bottom Ekman layer, it is not difficult to conclude that the anticyclonic eddy creates downwelling and suppresses upwelling in its centre, whereas the cyclonic eddy gradually lifts deeper water to the surface. Hence, headlands can operate as agents of baroclinic eddy formation which locally either suppress or enhance upwelling.

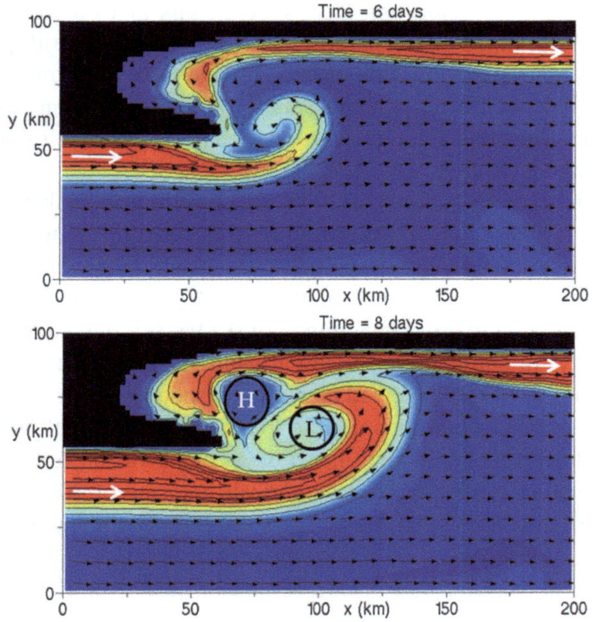

Fig. 5.16 Exercise 23. Snapshot surface distributions of tracer concentration (*color shading*) and flow field (*arrows*, averaged over 4 × 4 grid cells. *Red shading* refers to dense subsurface water being upwelled to the surface. *Bottom panel*: The *circle* marked "H" (for high-pressure centre) indicates a cyclonic eddy that operates to suppress upwelling. The *circle* marked "L" (for low-pressure centre) refers to an anticyclonic eddy

5.6.4 Additional Exercise for the Reader

Repeat this exercise with an overall onshore (northward) wind stress. Does the east facing coast of the bay support coastal upwelling?

5.6.5 Time-Splitting Methods

The reader will have noticed by now that three-dimensional model simulation are fairly slow and take a long time to complete. The reason of this time-step compliance with the CFL criterion for fast propagating surface gravity waves (Eq. 5.15). Enabling the rigid-lid condition (see Sect. 3.7) seems a solution, but this could lead to a bias in the dynamics predicted. Another method is to use different numerical time steps in different parts of the model code, a method called *time splitting*. To this end, small time steps only need to be applied within the S.O.R. iteration involving the surface pressure field, whereas all other parts such as advective and diffusive changes can employ a much larger time step. Note that the latter still needs to satisfy the stability criterion Eq. (5.14).

The first step of the time-stepping procedure is to predict the density field and first-guess changes of the velocity field, Δu, Δv and Δw, using an "internal" time step Δt_i being a multiple of the "external" time step Δt_e; that is,

$$\Delta t_i = m\Delta t_e$$

where the multiplier m is a positive integer. In a second step and prior to the S.O.R. iteration, the vector (u^*, v^*, w^*) is initialised with the current velocity field (u^n, v^n, w^n). After this, the S.O.R. iteration is repeated m times whereby the first-guess velocity is updated using equal fractions of the first-guess velocity change:

$$\begin{aligned} u^* &\leftarrow u^* + \Delta u/m \\ v^* &\leftarrow v^* + \Delta v/m \\ w^* &\leftarrow w^* + \Delta w/m \end{aligned}$$

An important rule when using a time-splitting method is that the internal time step should be not more than about 20 times the external time step, provided that all stability criteria are met. Otherwise, the external and internal modes inherent with the dynamics could drift apart from each other leading to a bias in the predictions. The author has tested the time-splitting method for this exercise achieving a three-fold reduction of the total simulation time. A modified code for Exercise 23 using the time-splitting method is contained in the folder "Miscellaneous/Time Splitting for Exercise 23" on the book's ftp site. The following exercises, however, do not make use of this method.

5.7 The Thermohaline Circulation

5.7.1 The Abyssal Circulation

The abyssal circulation of the ocean is far less well known than the wind-driven surface circulation. Knowledge about the deep circulation of the oceans has been derived from spatial distributions of chemical tracers (Broecker and Peng, 1982) and water mass properties; that is, temperature and salinity. Water-mass analysis is a technique that allows to determine the relative distributions of distinct source water masses along their path in the deep ocean. Tomczak and Godfrey (2003) give a general overview of the method of water-mass analysis.

The abyssal circulation, also called *thermohaline circulation*, is driven by convective formation of dense water in particular regions of the world ocean. The key players in this deep and bottom circulation are North Atlantic Deep Water (abbreviated as NADW), formed in the Labrador and Greenland Seas in the northern North Atlantic Ocean, and Antarctic Bottom Water, formed in the Weddell Sea and the Ross Sea of Antarctica. Influenced by the Coriolis force, dense water leaving these regions appears as bottom-arrested gravity plumes.

The lower limb or cold branch of the thermohaline circulation in the north Atlantic Ocean is the so-called *Deep Western Boundary Current* (abbreviated as DWBC). It moves dense water formed by open-ocean convection in the Labrador and Greenland Seas southward as a narrow subsurface flow along the western continental slope. The Deep Western Boundary Current can cross the equator and some of its water eventually merges with the Antarctic Circumpolar Current to flow around Antarctica and into the Indian and Pacific Oceans as North Atlantic Deep Water (NADW). The pathway of NADW is associated with a time scale of 1,500 years, which can been derived from radiocarbon dating, a method first described by Arnold and Libby (1949).

In order to conserve volume, southward spreading of the DWBC must be accompanied by a northward return flow. If the volume transport associated with the DWBC increases during its travel owing to entrainment of ambient water, the volume transport of the return flow has to increase as well. Open scientific questions are where this return flow occurs and what the forcing mechanisms of this return flow are. These questions are difficult to answer given that deep-ocean currents are generally very weak and difficult to measure and given that currents of the upper 1,000–2,000 m of the water column are predominantly driven by winds which overshadows other flow contributions.

5.7.2 The Stommel-Arons Model

The permanent thermocline is a zone of rapid decrease of temperature with depth in the tropics and subtropics reaching from the base of the seasonal thermocline to a depth of 1,500–2,000 m. In the sole presence of vertical diffusion of heat, the permanent thermocline would gradually deepen and the abyssal oceans would gradually become warmer until a thermodynamic equilibrium is being established. Since this does not happen, there must be another process that overrides the downward diffusive heat flux.

Based on earlier suggestions of Stommel (1958), Stommel and Arons (1960) proposed a model for the abyssal circulation of the world ocean that is still subject to intense debate by the scientific community. A key ingredient of this analytical model, commonly called the *Stommel-Arons model*, is the assumption that large-scale upwelling induced by geostrophic flow in the abyssal ocean balances vertical diffusion of heat, such that the structure of the permanent thermocline is maintained. Consider a two-layer ocean for illustration of this mechanism. For simplicity, the deep layer is assumed to have uniform thickness h and density. Using the beta-plane approximation (Eq. 4.7) gives a relationship between the vertical speed at the top of bottom layer and the meridional geostrophic flow component v that can be written as:

$$w = \frac{\beta h}{f} v \tag{5.16}$$

Consequently, upwelling at the top of the bottom layer is associated with a poleward geostrophic flow in this layer. A meridional flow of a speed of $v = 1$ mm/s with $h = 2{,}000$ m, for instance, would create a vertical upward displacement of isotherms at a rate of 14 m per year.

A further ingredient of the Stommel-Arons model is the prescription of two separate regions of dense-water formation that operate as volume sources for the deep layer; that is, the Greenland Sea/Labrador Sea and the Weddell Sea (ignoring the Ross Sea contribution to Antarctic Bottom Water formation). Volume conservation implies that poleward flow in deeper layers of the north Atlantic Ocean must be returned southward by a narrow DWBC. Nevertheless, the volume carried by this boundary current must be greater than that introduced at the source region given that it also has to return the volume inherent with the poleward flow. To this end, the DWBC carries some surplus volume across the equator and into the south Atlantic Ocean.

Figure 5.17 illustrates the analytical result of the Stommel-Arons Model for the Atlantic Ocean. Several inconsistencies of this model have been identified in recent years. These include: (a) observed deep-ocean values of eddy diffusivity are much less than those required to reproduce the Western Boundary Current with the Stommel-Arons model; (b) the model does not include any dynamics actually driving poleward deep flows, (c) variable bathymetry, such as mid-ocean ridges, are not included, despite their control on the possible pathways of deep flows. Despite these shortcomings, the Stommel-Arons model, proposed half a century ago, is still the backbone of many present-day research projects undertaken in the deep sea.

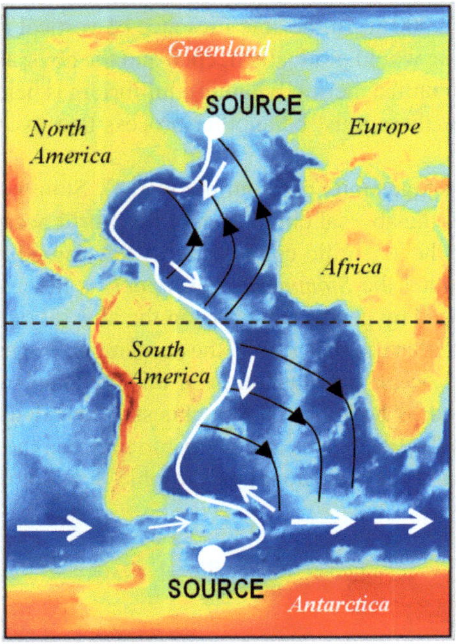

Fig. 5.17 Schematic of the deep-ocean circulation in the Atlantic Ocean after Stommel (1958) using GEBCO (General Bathymetric Chart of the Oceans) data

5.8 Exercise 24: The Abyssal Circulation

5.8.1 Aim

The aim of this exercise is to explore the dynamics of the deep circulation establishing in a closed ocean basin on the beta plane.

5.8.2 Task Description

We consider a model domain of 1,000 × 1,000 km in horizontal extent (Fig. 5.18), resolved by lateral grid spacings of $\Delta x = \Delta y = 20$ km, and a constant total depth of 1,000 m, resolved by a vertical grid spacing of $\Delta z = 200$ m. The ocean is initially at rest and uniform in density with $\rho = 1{,}028$ km/m^3. The model domain includes a separate semi-enclosed region of 200 × 200 km in horizontal dimension mimicking a region of dense-water formation such as the Greenland Sea or the Labrador Sea in the real situation. The model is forced by prescription of a density flux of 1 kg/m^3 per 30 days applied to the entire water column north of $y = 880$ km. This forcing is maintained during the entire model simulation lasting 120 days with data outputs on a daily basis.

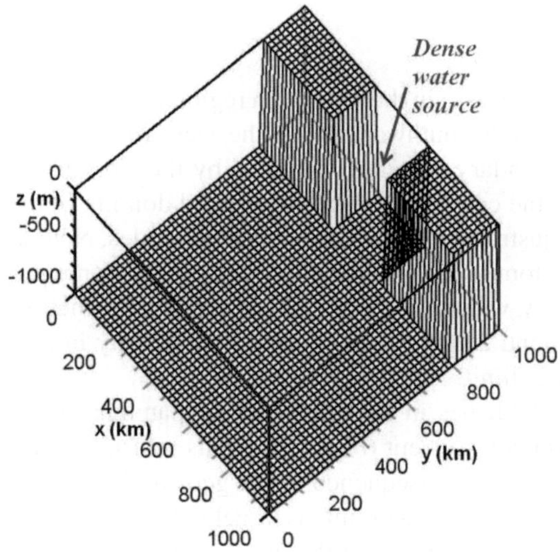

Fig. 5.18 Model domain for Exercise 24

Lateral eddy viscosity and eddy diffusivity are set to a uniform value of $A_h = K_h = 1\,\mathrm{m^2/s}$ using no-slip conditions for flow parallel to coastlines. Vertical eddy diffusivities and eddy viscosities are diagnosed from Kochergin's turbulence closure scheme. The bottom friction parameter is set to a value of $r = 0.001$. The beta-plane approximation assumes a Coriolis parameter that varies with meridional distance y according to:

$$f(y) = f_o + \beta\, y \tag{5.17}$$

In this exercise, we use a reference Coriolis parameter of $f_o = 1 \times 10^{-4}\,\mathrm{s^{-1}}$ (mid-latitudes in Northern Hemisphere) and a meridional variation of the Coriolis parameter of $\beta = 2.2 \times 10^{-11}\,\mathrm{s^{-1}\,m^{-1}}$. A total of 5,000 non-buoyant floats are initiated at random locations of the model domain to trace the trajectories of water parcels (see Sect. 3.16). In addition to this, Eulerian tracer concentration is used to identify regions experiencing upwelling during the simulation. To this end, the initial tracer concentration is set to zero in the upper 500 m of the water column and to values of unity below.

Owing to the smallness of resultant vertical speeds and associated round-off errors, predicted concentration fields in the author's pilot experiments showed some unwanted noisy patterns. To hide this, lateral eddy diffusivity used in the advection-diffusion equation for tracer concentration is increased to $100\,\mathrm{m^2/s}$ as a means of smoothing. This has no dynamical implications. The time step is set to $\Delta t = 200\,\mathrm{s}$ using the rigid-lid approximation to eliminate fast propagating surface gravity waves. The pressure accuracy of the S.O.R. iteration is set to $\epsilon = 0.01\,\mathrm{Pa}$.

5.8.3 Results

The gradual density increase in the formation region creates a Deep Western Boundary Current that spreads southward along the western side of the model domain (Fig. 5.19). This boundary current is initiated by the Coriolis force which pushes deep water against the coast. The resultant upward doming of the density interface and geostrophic adjustment lead to steady-state dynamics. Note that an anticyclonic eddy appears in bottom layers of the dense-water formation region. Owing to a net influx of low-density water, the density excess in the formation region reaches a steady value of around 2 kg/m^3. This implies that density in a formation region is bounded by an upper limit.

A return flow establishes in the upper ocean spanning the entire length of the Deep Western Boundary Current (Fig. 5.20). This return flow is created by a drop of the coastal sea level as consequence of the geostrophic adjustment of the dense bottom layer. The adjacent ocean remains largely at rest. A counterclockwise rotating eddy appears in surface layers of the dense-water formation region. Upwelling along the path of the boundary current leads to entrainment of bottom water into the surface return flow. There is some indication of meandering of the boundary current, but the inherent dynamics are grossly biased by the coarse lateral grid spacing chosen.

In contrast to the Stommel-Arons model, a vertical rather than horizontal overturning circulation establishes here being concentrated to a narrow zone along the western boundary. The final locations of selected Lagrangian floats nicely reveal the boundary current (Fig. 5.21) together with the return flow that establishes in the surface ocean (Fig. 5.22). This model application does not create any return flows in

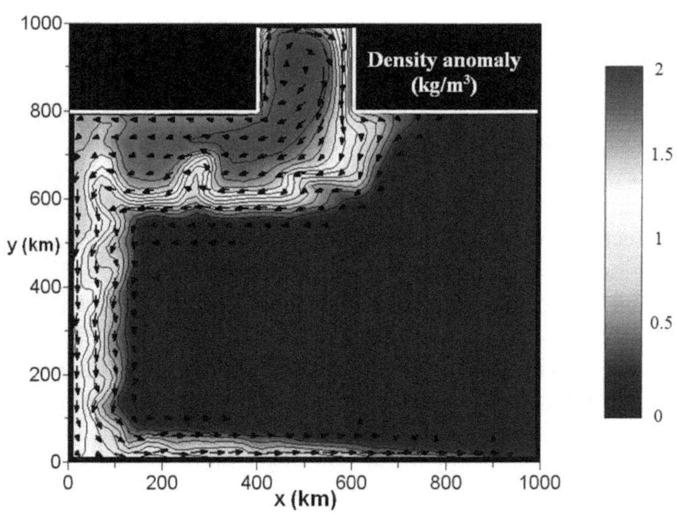

Fig. 5.19 Exercise 24. Distributions of density (*shading* and *contours*) and horizontal flow field (*arrows*) in the lowermost 200 m of the water column domain after 120 days of simulation. *Arrows* for speeds <1 cm/s are omitted. Velocities are averaged over 2 × 2 grid cells

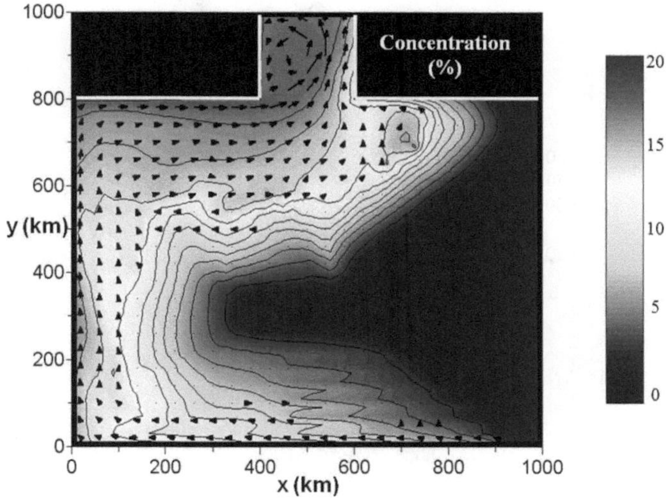

Fig. 5.20 Exercise 24. Distributions of tracer concentration (*shading* and *contours*) and horizontal flow field (*arrows*) in the uppermost 200 m of the water column after 120 days of simulation. Arrows for speeds <1 cm/s are omitted. Velocities are averaged over 2 × 2 grid cells

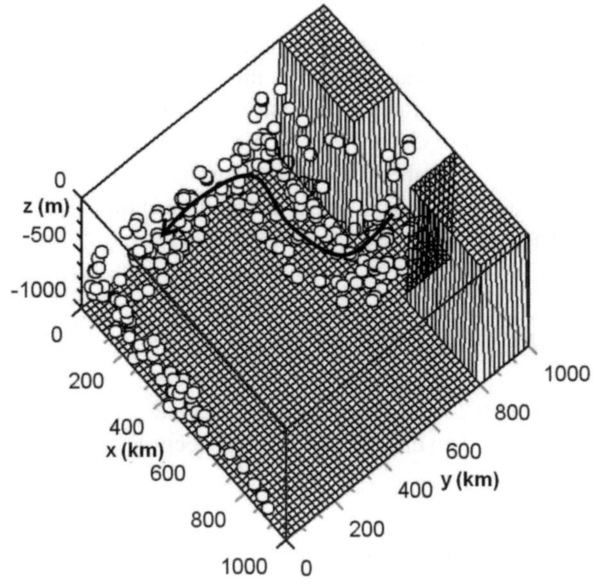

Fig. 5.21 Exercise 24. Locations of Lagrangian floats, initially located in the formation region of dense water north of $y = 800$ km after 120 days of simulation

the ocean interior, presumably because of the relatively short simulation time. Note that the artificial southern coastline supports an eastward zonal flow. In reality, the Deep Western Boundary Current encounters the equator. Dynamical consequences of this encounter are discussed in the next section.

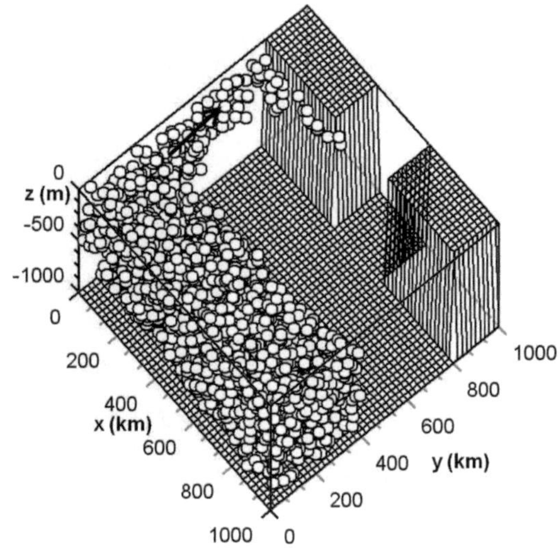

Fig. 5.22 Exercise 24. Locations of Lagrangian floats, initially located south of $y = 350$ km, after 120 days of simulation

Findings presented here show an intrinsic coupling between the deep and surface flows in the western boundary region of the Atlantic Ocean. This suggests that the density-driven DWBC operates to intensify the wind-driven Western Boundary Current in the north Atlantic Ocean; that is, the Gulf Stream and its extension, the North Atlantic Current. A number of recent research projects explore this coupling mechanism which is deemed to have a substantial impact on the climate of northern Europe on time scales of 10–20 years (Broecker, 1999).

5.8.4 Additional Exercise for the Reader

Repeat this exercise with $\beta = 0$, yielding the so-called f plane. Does the β-effect have a noticeable influence on the dynamics of the Deep Western Boundary Current?

5.8.5 Improved Float Tracking

Float locations are defined as distances from the point of origin of the Cartesian coordinate system. In this book, the point of origin is defined at the vertical velocity grid point w at $k = 1$, $j = 1$ and $i = 1$, which is aligned with the sea surface. Prediction of the displacement of a float requires information of velocity in adjacent grid points. To this end, the grid cell containing the float need to be identified and the rest follows. The procedure is demonstrated here using the x-direction only. Analog recipes apply for the other velocity components, noting that treatment for the vertical

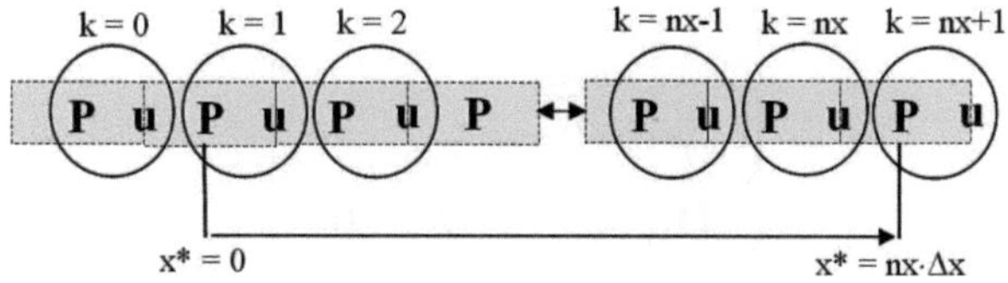

Fig. 5.23 Illustration showing the grid configuration and float coordinate for the x-direction. *Circles* group grid points belonging to a certain grid index

direction is slightly different from that of horizontal directions. Figure 5.23 shows the grid configuration together with the float's x-coordinate. The first step of the procedure is to locate the grid cell containing a float with reference to the closest scalar grid point. This can be done with:

$$\text{kpos} = \text{INT}\left(\frac{x^*}{\Delta x} + 0.5\right) + 1$$

where x^* is the actual x-coordinate of the float, and "INT" truncates real numbers into full (integer) numbers. The simplest scheme would then be to average surrounding velocity grid points onto the scalar grid point, yielding a velocity of:

$$\langle u \rangle = 0.5(u_w + u_e)$$

where $u_w = u(\text{kpos} - 1)$ and $u_e = u(\text{kpos})$. This velocity can be used to move the float around on the basis of the simple displacement equation:

$$\frac{dx^*}{dt} = \langle u \rangle$$

This scheme has been used in some of the previous exercises, noting that stranding of floats in dry grid cells can be avoided with use of additional conditions. An improved scheme interpolates (rather than averages) the velocity from surrounding grid points onto the location of the float. In the x-direction, for instance, this interpolation leads to:

$$u(x^*) = u_w + \delta x^* \frac{u_e - u_w}{\Delta x} \tag{5.18}$$

where δx^* is the distance from the float to the western cell face (Fig. 5.24), given by:

$$\delta x^* = x^* - \Delta x\,(\text{kpos} - 1) + 0.5\Delta x = x^* + (1.5 - \text{kpos})\Delta x$$

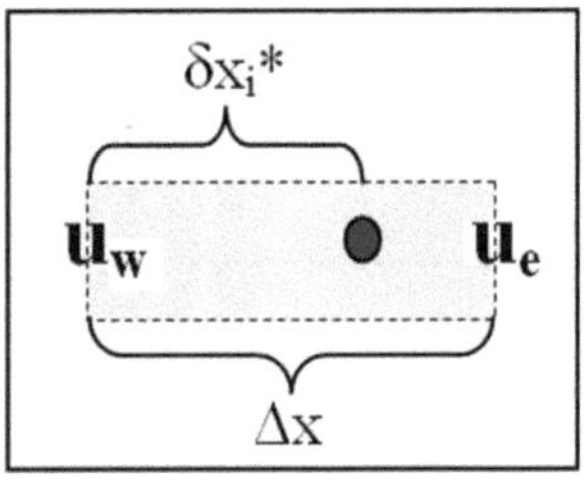

Fig. 5.24 Definition of the distance δx^*. The *circle* indicates the float location. The subscript "i" refers to the start location of the float

The displacement equation is now given by:

$$\frac{dx^*}{dt} = u(x^*) \tag{5.19}$$

With the abbreviation $u'_x = (u_e - u_w)(\Delta x)$, Eqs. (5.18) and (5.19) can be combined and integrated to yield:

$$\int_t^{t+\Delta t} dt = \int_i^e \frac{dx^*}{u(x^*)}$$

where the integral boundaries "i" and "e" refer to the start and end locations of a float. This integral gives:

$$u'_x \Delta t = \ln\left(\frac{u_w + \delta x_e^* u'_x}{u_w + \delta x_i^* u'_x}\right)$$

Finally, the distance traveled by the float over a time span of Δt can then be calculated from:

$$\Delta x^* = x_e^* - x_i^* = \left(\frac{u_w}{u'_x} + \delta x_i^*\right)\left[\exp(u'_x \Delta t) - 1\right] \tag{5.20}$$

Notice that this scheme collapses when u'_x approaches zero. In this case, the simple averaging method must be used instead. Errors arise when a float crosses a cell boundary during a time step. This can be accommodated in the code by splitting the method into parts, whereby the first sub-time step is based on the time is takes for a float to approach the eastern or western cell face, and the remaining time step continues with the adjacent scalar grid point as a new cell centre. To this end, the first sub-time step can be calculated from:

$$\Delta t_{\text{min}} = \frac{1}{u'_x} \ln \left(\frac{\delta x_{\text{edge}} u'_x}{u_w + \delta x_i^* u'_x} + 1 \right)$$

In distance to the respective boundary δx_{edge} is defined as $\delta x_{\text{edge}} = \Delta x - \delta x_i^*$ for eastward flow and as $\delta x_{\text{edge}} = -\delta x_i^*$ for westward flow.

For the case of Lagrangian floats, turbulence is described by the probability that a particle is shifted a certain distance within a given time step (Maier-Reimer and Sündermann, 1982). After Maier-Reimer (1980), maximum diffusive velocities can be related to eddy diffusion coefficients. In the x-direction, for instance, this relationship reads:

$$u_{\text{max}} = \sqrt{\frac{6A_h}{\Delta t}}$$

Turbulent velocity fluctuations can then be determined for each float with the *Monte-Carlo method* which consists of multiplying each maximum current component by a random generated number between −1 and 1. A displacement equation of the form of Eq. (5.19) can be adopted to include turbulence effects on float motions.

Computer codes of this book only use the simple averaging method for float predictions and ignore effects of turbulence. The implementation of more advanced float tracking schemes remains for the reader.

5.9 The Equatorial Barrier

5.9.1 Inertial Oscillations About the Equator

A flow crossing the equator experiences a change of sign of the Coriolis force and therefore becomes deflected back towards the equator (Fig. 5.25). It then again overshoots the equator owing to its inertia, and the Coriolis force on the other side of the equator moves it back, and so on, in an oscillatory fashion. This unique behavior, in which the equator acts as a waveguide, is the signature of inertial oscillations about the equator. Whereas inertia oscillations in off-equatorial regions are associated with closed flow trajectories (in the absence of ambient flow), those centred on the equator are always associated with a net eastward flow. The period of equatorial inertial oscillations is approximately given by (e.g., Cushman-Roisin, 1994):

$$T_{\text{eq}} = \frac{1}{\sqrt{\beta V_o}} \tag{5.21}$$

where β is the meridional variation of the Coriolis parameter at the equator ($\beta \approx 2.28 \times 10^{-11}$ m^{-1} s^{1}), and V_o is the initial speed of the flow crossing the equator. With $V_o = 0.2$ m/s, Eq. (5.21) yields 5.4 days. The radius of the half-circle paths of

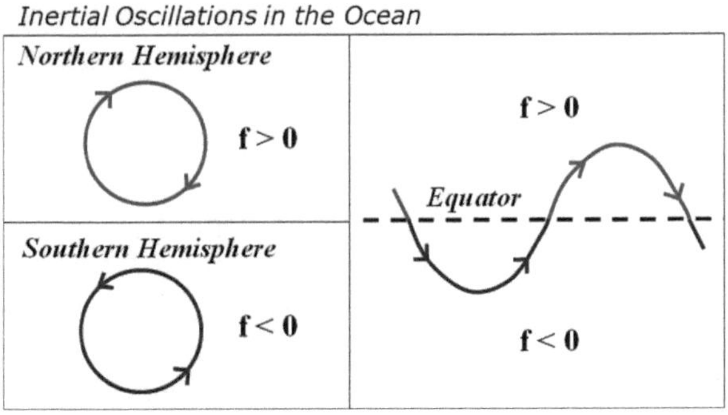

Fig. 5.25 Schematic of inertial oscillations in the ocean

the flow can be estimated at:

$$r_{\mathrm{eq}} = V_o \, T_{\mathrm{eq}} = \sqrt{\frac{V_o}{\beta}} \tag{5.22}$$

For the previous values, this is about 94 km, or almost 1° of geographical latitude. The length scale r_{eq} can be referred to as *equatorial inertial radius*.

5.9.2 Variation to Exercise 24

The interaction of a bottom-arrested Deep Western Boundary Current with the equator can be explored with a slight modification of the configuration of Exercise 24. The only change required is to modify the Coriolis parameter such that the equator, defined by $f_o = 0$ in Eq. (5.17), is shifted to y = 400 km of the previous configuration (see Fig. 5.22) and to enhance the beta effect by choosing $\beta = 2.2 \times 10^{-10}$ $\mathrm{m}^{-1}\ \mathrm{s}^{-1}$. The choice of $f_o = 0$ is commonly referred to as *equatorial beta-plane approximation*.

5.9.3 Results

As can be seen with the result (Fig. 5.26), the DWBC becomes subject to equatorial inertial oscillations deflecting this flow eastward along the equator. This is consistent with observational evidence (Bourlès et al., 2003). The simulated boundary current attains a speed of 0.9 m/s (real flows are much weaker than this). For this value, Eq. (5.10) gives an equatorial inertial radius of 64 km, which is not well resolved by the coarse mode grid chosen. Nevertheless, damped inertial oscillations are clearly visible in the prediction. Findings presented here suggest that the equator can operate as a barrier for the DWBC.

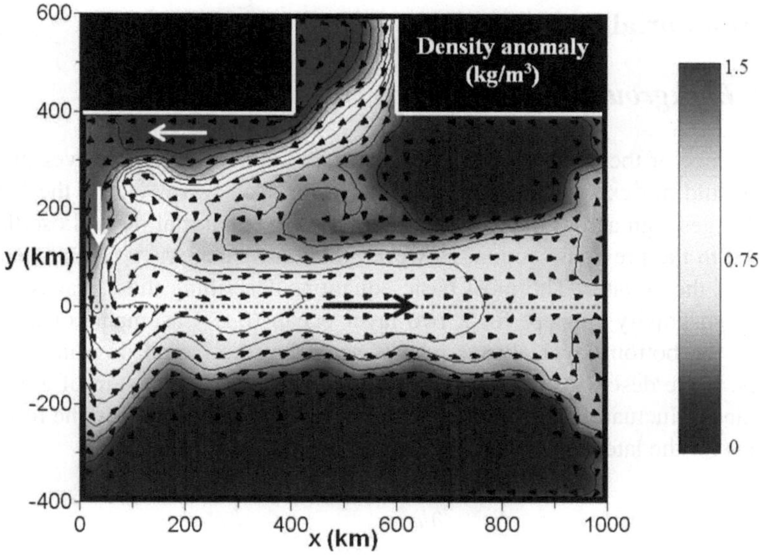

Fig. 5.26 Variation to Exercise 24. Distributions of density (*shading* and *contours*) and horizontal flow field (*arrows*) in the lowermost 200 m of the water column domain after 120 days of simulation. *Arrows* with speeds <1 cm/s are omitted. Velocities are averaged over 2×2 grid cells. The *dotted line* highlights the equator

Recent studies (Dengler *et al.*, 2004) indicate that the DWBC is a continuous flow across the equator to a geographical latitude of 8° S where it breaks up into eddies. This break-up into eddies is presumably initiated by strong curvature of the flow path that establishes south of the equator owing to the sign change of the Coriolis force (see Fig. 5.26). Interaction of the flow with variable topography, not considered here, might also contribute to eddy shedding.

5.9.4 Additional Exercise for the Reader

Repeat this exercise with an initially stratified ocean. To this end, increase the initial density of the lower 200 m of the water column by 2 kg/m^3 and apply the density forcing in the region of dense-water formation exclusively to the upper 800 m. Compare the resultant horizontal density fields for the lowest and the second lowest model levels. Note that, after some time, the density forcing applied will create unstable density stratification that, in the real world, would induce convective stirring of the water column. This mixing can be parameterised in Kochergin's turbulence closure scheme via use of a locally increased value of vertical eddy diffusivity.

5.10 Equatorial Waves

5.10.1 Background

The existence of the equator gives rise to special kinds of oceanic waves that otherwise would not exist. The dynamical reason for such waves is that the Coriolis force changes sign across the equator, giving rise to equatorial inertial oscillations described in the previous section. However, there are other wave types existing in vicinity of the equator. The most basic equations describing these waves rely on the reduced-gravity concept for a two-layer ocean (e.g. Cushman-Roisin, 1994) in which the bottom layer always adjusts such that the lateral flow in this layer vanishes. If we describe the thickness of the upper layer as the sum of a constant part h_o and a fluctuating part η^*, the reduced-gravity concept leads to the following equations for the lateral currents in the upper layer:

$$\begin{aligned} \frac{\partial u}{\partial t} - \beta\, y\, v &= -g' \frac{\partial \eta^*}{\partial x} \\ \frac{\partial v}{\partial t} + \beta\, y\, u &= -g' \frac{\partial \eta^*}{\partial y} \\ \frac{\partial \eta^*}{\partial t} + h_o \left(\frac{\partial u}{\partial x} + \frac{\partial v}{\partial y} \right) &= 0 \end{aligned} \tag{5.23}$$

where g' is reduced gravity. The equatorial beta-plane approximation has been used here together with the assumption that variations of the top-layer thickness remain small, so that a constant thickness h_o can be used in the volume-conservation equation. Note that η^* is the negative of the interface displacement, simply because interface displacements and layerthickness changes are opposite to each other.

5.10.2 Equatorial Kelvin Waves

The first breed of equatorial waves to be discussed are *equatorial Kelvin waves* that can be extracted from the above equations with the assumption of vanishing meridional flow; that is, $v = 0$, everywhere in the domain. In this case, the above equations can be written as:

$$\begin{aligned} \frac{\partial u}{\partial t} &= -g' \frac{\partial \eta^*}{\partial x} \\ \beta\, y\, u &= -g' \frac{\partial \eta^*}{\partial y} \\ \frac{\partial \eta^*}{\partial t} + h_o \frac{\partial u}{\partial x} &= 0 \end{aligned} \tag{5.24}$$

The solution for zonally propagating waves is given by:

$$
\begin{aligned}
u &= u_o \exp\left(-\frac{y^2}{2R_{\text{eq}}^2}\right)\cos(kx-\omega t)\\
v &= 0\\
\eta^* &= A_o \exp\left(-\frac{y^2}{2R_{\text{eq}}^2}\right)\cos(kx-\omega t)
\end{aligned}
\tag{5.25}
$$

where u_o and A_o are interrelated constants, the wave number is defined by $k = 2\pi/L$ with L being wavelength, and the wave frequency is defined by $\omega = 2\pi/T$ with T being wave period. The so-called *equatorial radius of deformation* is hereby defined as:

$$
R_{\text{eq}} = \sqrt{\frac{\sqrt{g'h_o}}{\beta}}
\tag{5.26}
$$

It can be shown that these waves propagate at the speed of long internal gravity waves, $\sqrt{g'h_o}$, and that they can only propagate eastward along the equator. The dynamic similarity to coastal Kelvin Waves justifies their name. The phase speed of such waves is 0.5–1 m/s and the trapping distance, estimated by R_{eq}, is about 50–200 km.

5.10.3 Other Equatorially Trapped Waves

The following section has been adopted from Cushman-Roisin (1994). We seek solutions of Eq. (5.24) that describe waves propagating in the zonal direction, but this time in a more generalised form of:

$$
\begin{aligned}
u &= U(y)\cos(kx-\omega t)\\
v &= V(y)\sin(kx-\omega t)\\
\eta^* &= A(y)\cos(kx-\omega t)
\end{aligned}
\tag{5.27}
$$

where the amplitude functions $U(y)$, $V(y)$ and $A(y)$ need to be determined. Insertion of this approach in Eq. (5.23) leads to the following relationships between the amplitude functions:

$$
\begin{aligned}
\omega U - \beta\, y\, V &= g'kA\\
\omega V - \beta\, y\, U &= g'\, dA/dy\\
\omega A - h_o kU &= -h_o\, dV/dy
\end{aligned}
\tag{5.28}
$$

Combinations of the latter equations yield a single equation governing the meridional structure of $V(y)$:

$$\frac{\partial^2 V}{\partial y^2} + \left(\frac{\omega^2 - \beta^2 y^2}{g' h_o} - \frac{\beta k}{\omega} - k^2\right) V = 0 \tag{5.29}$$

It can be shown that this equation has the solutions:

$$V(y) = H_n \left(\frac{y}{R_{\text{eq}}}\right) \exp\left(-\frac{y^2}{2R_{\text{eq}}^2}\right) \tag{5.30}$$

where H_n is a so-called *Hermite polynomial* of the order of n with the first modes being given by:

$$H_0(\psi) = 1$$
$$H_1(\psi) = 2\psi$$
$$H_2(\psi) = 4\psi^2 - 2$$

Notice that even polynomials are symmetric about the equator, whereas those of odd order are antisymmetric. All waves are trapped in vicinity of the equator on a trapping distance given by the equatorial radius of deformation and their dispersion relation is given by:

$$\frac{\omega^2}{g' h_o} - k^2 - \frac{\beta k}{\omega} = \frac{(2n+1)\beta}{\sqrt{g' h_o}} \tag{5.31}$$

where n is a positive integer including zero. Accordingly, waves are composed of a discrete set of modes. For $n \geq 1$, the waves subdivide into two classes. One branch of waves are relatively fast propagating *equatorially trapped inertia-gravity waves* that follow a dispersion relation according to:

$$\omega \approx \sqrt{\frac{2n+1}{T_{\text{eq}}^2} + k^2 g' h_o} \tag{5.32}$$

where the so-called *equatorial inertial period* is defined by:

$$T_{\text{eq}} = \sqrt{\frac{1}{\beta \sqrt{g' h_o}}} \tag{5.33}$$

Typical values for T_{eq} are 2–3 days. This branch of equatorially trapped waves includes a modified form of equatorial Kelvin Waves that, unlike the wave solution described in the previous section, involves nonzero meridional flow. Figure 5.27 illustrates the surface pressure field and currents of such waves that, again, can only propagate eastward along the equator.

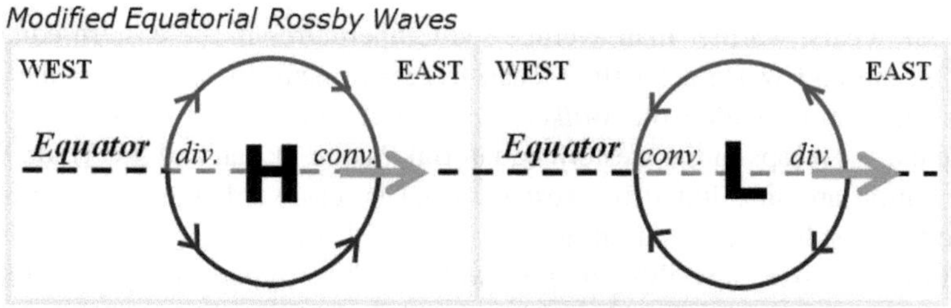

Fig. 5.27 Illustration of the horizontal flow field associated with centres of high pressure (elevated sea level) or low pressure (lowered sea level) at the equator. Both patterns create flows that trigger eastward propagation of the disturbance

The other branch of equatorially trapped waves follows the dispersion relation:

$$\omega \approx -\frac{\beta k}{k^2 + (2n+1)/R_{eq}^2} \tag{5.34}$$

These waves, called *equatorial planetary waves* or *equatorial Rossby waves*, travel westward at slow speeds and, hence, opposite to equatorial Kelvin waves. The special case $n = 0$ is characterised by the dispersion relation:

$$\omega T_{eq} - \frac{1}{\omega T_{eq}} = k R_{eq} \tag{5.35}$$

This wave exhibits a mixed behavior between planetary and inertia-gravity waves and is therefore called *mixed planetary-inertia-gravity wave*. Interestingly, unlike equatorial Kelvin waves, the mixed wave propagates westward along the equator.

Although all these distinct equatorial wave types exist in theory, they appear in the real ocean only if being excited by some external mechanism. For instance, dense water flows passing the equator can trigger equatorial inertial oscillations (see previous exercise) and disturbances of the larger-scale equatorial wind system can trigger equatorial Rossby waves (see next exercise). Spatial and temporal scales of the initial perturbation determine hereby the wave type and wave mode that, in the end, appears magnified in the ocean.

5.11 The El-Niño Southern Oscillation

5.11.1 Background

The El-Niño Southern Oscillation (ENSO) is a synonym of a pronounced climate variation that occurs in the equatorial and tropical Pacific region on time scales of 2–10 years. This oscillation is caused by a large-scale dynamic and thermodynamic interaction between the ocean and the atmosphere. Only a brief description of this interaction is given here.

In some years, warmer than average sea temperatures in the eastern equatorial Indian Ocean and western Pacific Ocean trigger a local disturbance of the atmospheric circulation, called the *Walker circulation*. As a result of this, a tropical cyclone pair develops in the western equatorial Pacific Ocean and the Trade Winds in this region are substituted by westerly (that is, eastward blowing) winds. The modified atmospheric circulation, in turn, creates disturbances in the ocean's density field via localised upwelling or downwelling. This triggers an equatorial Kelvin wave propagating eastward to the other side of the Pacific where is suppresses the semipermanent coastal upwelling of cold and nutrient-enriched water off the Peruvian and Californian coasts and leads to widespread kills of plankton, fish, and sea birds. These occasions are called El Niño (Spanish for "the child") events for they occur usually around the Christmas season.

5.12 Exercise 25: Simulation of an El-Niño Event

5.12.1 Aim

The aim of this exercise is the simulate disturbances in the equatorial western Pacific Ocean leading to an El-Niño event on the eastern side of the Pacific Ocean.

5.12.2 Task Description

The model domain has a horizontal extent of 2,000 × 2,000 km, resolved by lateral grid spacings of $\Delta x = \Delta y = 20$ km (Fig. 5.28). Coasts are located in the west and in the east. The northern and southern limits of the model domain are treated as open boundaries. Total water depth is set to 250 m to allow for relatively long numerical time steps noting that the free-surface version of the model is being used. Vertical grid spacing is set to $\Delta z = 50$ m, and the numerical time step is set to 120 secs. Initially, the ocean is at rest. Seawater density in the upper 100 m of the water column is set to 1,027 kg/m^3. Underneath, the density is 1,028 kg/m^3. The phase speed of internal gravity waves associated with this density configuration is $c = 0.75$ m/s, using Eq. (3.63). The equatorial beta-plane approximation is employed with:

$$f = \beta y \tag{5.36}$$

where y is the meridional coordinate and β is set to $\beta = 2.5 \times 10^{-11}\,\mathrm{m}^{-1}\,\mathrm{s}^{-1}$. The equator at $y = 0$ cuts the model domain in half. The equatorial radius of deformation (take Eq. 5.27) is about 174 km, which is reasonably well resolved by the lateral grid spacing chosen. The equatorial inertial period (take Eq. 5.34) is approximately 2.7 days.

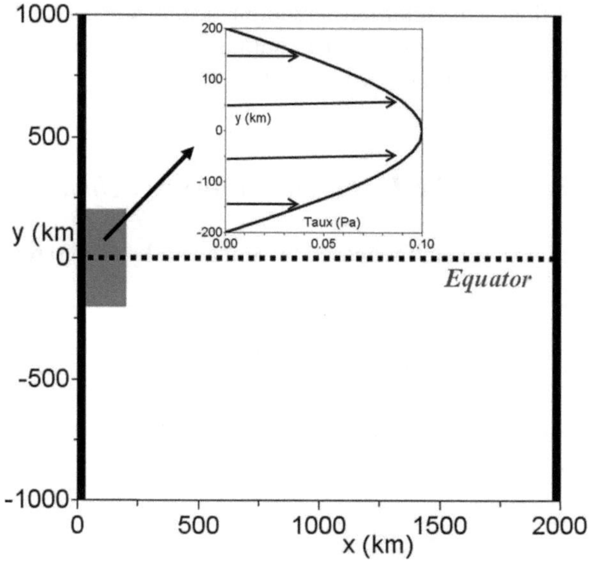

Fig. 5.28 Horizontal extent of the model domain and wind-stress forcing for Exercise 25

The model is forced via prescription of a short-lived westerly wind burst. This forcing consists of a eastward wind stress of cosine shape in the y-direction (see Fig. 5.26) of a maximum magnitude of 0.1 Pa at the equator. This wind-stress forcing is only applied within a distance of 200 km from the western coast. The wind-stress magnitude is gradually adjusted to its final value over the first 5 days of simulation and turned off afterwards. Modified cyclic boundary conditions are used for the northern and southern open boundaries via use of a "mirror" condition for the meridional flow component reading:

$$v(i, 0, k) = -v(i, ny, k)$$
$$v(i, ny + 1, k) = -v(i, 1, k)$$

The advanced turbulence closure scheme proposed by Smagorinsky (1963) is implemented to parameterise lateral eddy viscosity and diffusivity. This scheme, described in the next section, is frequently employed by hydrodynamic modellers. The free parameter of the scheme is set to $c_1 = 0.1$, and a uniform background value of 5 m^2/s is added. The Kochergin turbulence closure scheme is employed again to describe the effect of vertical mixing using $c_2 = 0.15$ and a wind-induced value of $K_z = A_z = 0.05\,m^2/s$ in the upper 100 m of the water column.

Both Eulerian tracer concentration and Lagrangian floats are used for visualisation of the dynamics. The tracer concentration field is initialised with zero values in the upper 100 m of the water column and values of unity underneath. A total of

5,000 Lagrangian floats are initialised at random locations of the model domain. The total simulation time is 50 days with data outputs at 12-hrs intervals. The pressure accuracy of the S.O.R. iteration is set to $\epsilon = 0.01$ Pa.

5.12.3 The Smagorinsky Turbulence Closure Scheme

The turbulence closure scheme proposed by Smagorinsky (1963) reads:

$$A_h = c_1 \Delta x \Delta y \sqrt{\left(\frac{\partial u}{\partial x}\right)^2 + \left(\frac{\partial v}{\partial y}\right)^2 + 0.5\left(\frac{\partial u}{\partial y} + \frac{\partial v}{\partial x}\right)^2} \tag{5.37}$$

where c_1 is a parameter often set to values between 0.1 and 0.2.

5.12.4 Warning

Completion of this simulation might take one or more days. Also, SciLab might crash occasionally owing to memory limitations. To avoid these problems, the reader might double lateral grid spacings to $\Delta x = \Delta y = 40$ km using half the number of grid points for both the x- and the y-direction. This, however, will incur a severe loss in spatial resolution.

5.12.5 Results

The westerly wind burst creates an initial depression of the thermocline travelling eastward along the equator over a distance of 2000 km in form of an equatorial Kelvin Wave (Fig. 5.29). The phase speed of this wave is approximately 66 cm/s, which is slightly less than predicted by theory (take Eq. 3.63). The imprint of the Kelvin Wave is a circular patch in the Eulerian tracer concentration field and, accordingly, in the density field. This imprint is associated with transient downward pumping of the pycnocline whereby the wave depresses the pycnocline by 30–50 m. The radius of the circular patch is 200 km, being of the order of the equatorial radius of deformation. A second wave develops behind the trailing wave.

As the trailing wave meets the eastern coastline, it disintegrates into two coastal Kelvin waves propagating away from the equator in the respective hemisphere. Turbulence-enhanced vertical mixing leaves behind some perturbation in the density field and weak currents in the western equatorial region of the model domain. With a closer inspection, we can also spot planetary Rossby waves not far from the western boundary slowly propagating westward.

Owing to the trade winds and the Coriolis parameter changing sign across the equator, equatorial regions are generally prone to upwelling which operates to

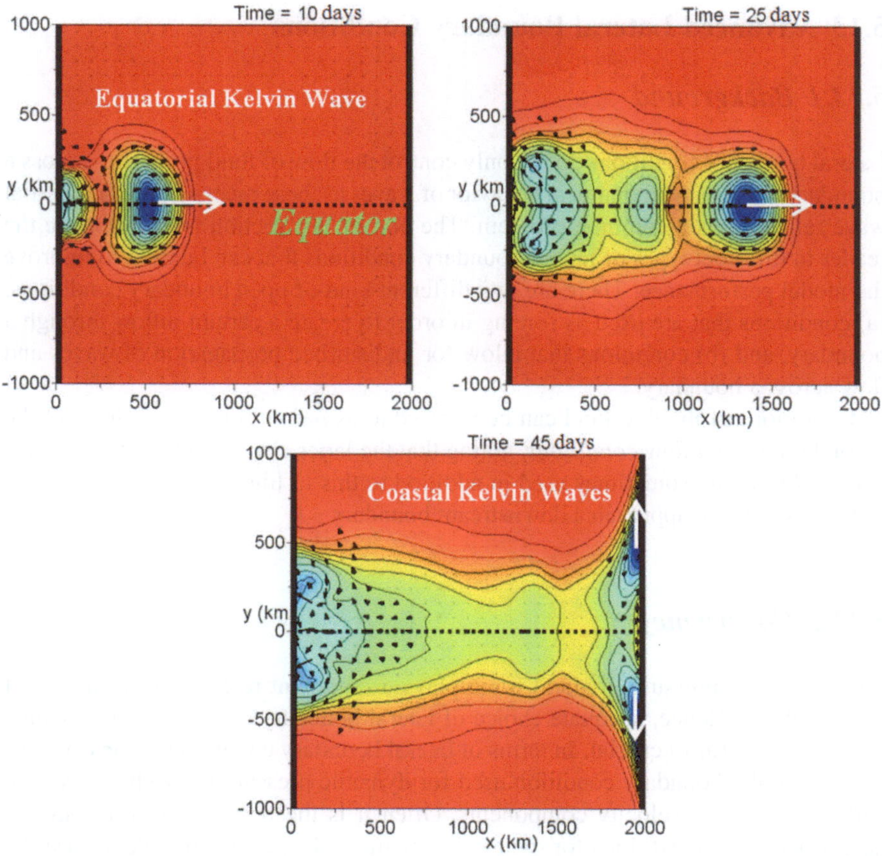

Fig. 5.29 Exercise 25. Selected snapshots of horizontal distributions of Eulerian tracer concentration (*color shading* and *contours*) at a depth horizon of 150 m. Blue shading refers to decreased concentrations. Black arrows are lateral flow vectors

lower sea-surface temperature by a couple of degrees. Transient equatorial Kelvin waves associated with a high-pressure centre (such as that simulated here) operate to suppress this equatorial upwelling along their path. Hence, satellite-derived sea-surface temperature can be used to trace the passage of equatorial Kelvin waves.

5.12.6 Additional Exercises for the Reader

Consider an easterly wind burst near the eastern coast of the model domain. Does any of the waves created propagate to the western side of the model domain?

5.13 Advanced Lateral Boundary Conditions

5.13.1 Background

Lateral boundary conditions do not only control the flow of fluid properties across a boundary but also the dynamical behavior of waves as these meet a boundary. Partial wave reflection is a common problem. The aim of this section is to introduce the reader to different types of lateral boundary conditions that can be used to improve the model performance. There are two different kind of open boundary conditions: (a) conditions that are used as forcing in order to create a certain inflow through a boundary, and (b) conditions that allow for undisturbed propagation of waves and flow across a boundary.

Conditions of the first kind can be referred to as *inflow conditions*, those of the second kind as *outflow conditions*, noting that the latter also includes wave signals. *Sponge layers* are sometimes used in addition to this to filter away dynamical disturbances as these approach a downstream boundary.

5.13.2 Consistency

The horizontal pressure gradient force plays a dominant role in the dynamics of oceanic flows. Hence, adequate choice of lateral boundary conditions for dynamic pressure is uttermost crucial. In terms of lateral boundary conditions, "consistency" means that the boundary condition used for dynamic pressure has to be consistent with those set for velocity components. Often it is the best approach to employ lateral boundary conditions for dynamic pressure only and, if possible, to use the numerical code to predict boundary values of the velocity component normal to the boundary. Prescription of boundary values for all variables is not recommended since this can lead to inconsistency in the dynamics and unwanted side effects.

5.13.3 Inflow Conditions

For wave problems excluding Coriolis effects, it is often sufficient to prescribe dynamic pressure variations at a lateral boundary and to calculate velocities inside the prediction code. Steady geostrophic inflows such as those in Exercise 21 are best realised via prescription of flow components normal to the boundary in conjunction with vanishing gradients of dynamic pressure normal to the boundary which filters away unwanted geostrophic flow running parallel to a boundary. Another commonly applied method is the method of one-way nesting of a smaller model domain inside a larger model domain and to use predictions from the larger domain as boundary conditions for the smaller domain.

Lateral boundaries for combined geostrophic flow and wave problems are difficult to deal with. One solution would be to decompose dynamic pressure into a time-mean component driving the geostrophic inflow and a superposed time-variable component representing incident waves such as tides. The geostrophic part can then be described along the boundary as a background distribution, gradually blended in during the first few days of a simulation, and the time-variable part can be added via a prognostic equation of the form:

$$\frac{\partial q_b}{\partial t} = q_o \omega \sin(\omega t) \tag{5.38}$$

where q_b is boundary dynamic pressure, and q_o and ω are amplitude and frequency of the time-variable forcing. The reader is encouraged to test this type of forcing for a simplified open channel configuration.

Inflow conditions for water properties such as temperature and salinity are straight forward. These properties are simply prescribed as fixed boundary values so that the inflow carries these via the advection scheme into the model domain. Establishment of sharp density gradients near open boundaries can be avoided with the use of an adjustment method, commonly called *Rayleigh damping*, of the form:

$$\psi_b^{n+1} = \psi_b^n + \frac{\Delta t}{T}\left(\psi_o - \psi_b^n\right) \tag{5.39}$$

where ψ_b is the boundary value of either temperature or salinity, T is a prescribed adjustment period, and ψ_o is the target boundary value.

5.13.4 Outflow Conditions

The formulation of outflow conditions at open boundaries is not a trivial task given that both steady currents and wave signals can simultaneously interfere with such a boundary. Unwanted partial wave reflection at open boundaries is a common problem. Different types of outflow boundary conditions are best demonstrated with a focus on the propagation of long linear surface gravity waves in a channel of uniform depth h_o. The dynamics of such waves can be approximated by the equations:

$$\frac{\partial u}{\partial t} = -g\frac{\partial \eta}{\partial x} \tag{5.40}$$

$$\frac{\partial \eta}{\partial t} = -h_o\frac{\partial u}{\partial x} \tag{5.41}$$

where u is velocity, t is time, g is acceleration due to gravity, η is seasurface elevation, and x is distance along the channel. Assumptions are that the wave period is short compared with the Coriolis force (so that the latter can be ignored), that the wave's phase speed exceeds the flow speed by far (so that the nonlinear terms can be ignored), that the hydrostatic balance holds (such that the resultant flow is

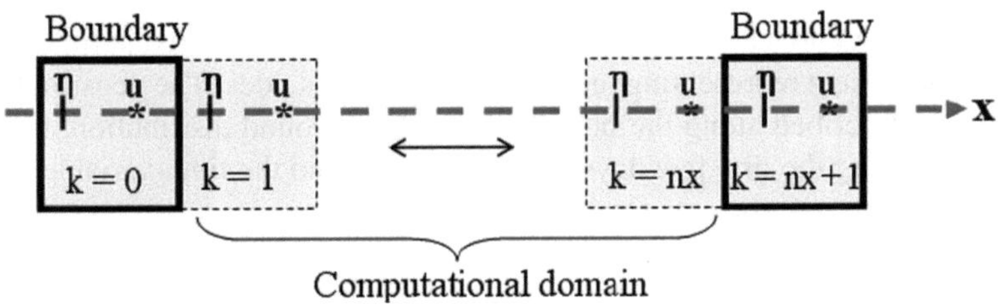

Fig. 5.30 Arakawa C-grid for one-dimensional shallow-water applications

barotropic), that frictional effects are negligibly small, and that sea-level anomalies η are small compared with undisturbed water depth h_o (leading to simplification of the volume conservation equation Eq. (5.41)).

On the basis of the one-dimensional version of the Arakawa C-grid (Fig. 5.30), the explicit numerical finite-difference scheme of the above equations can be formulated in three subsequent steps given by:

$$u_k^* = u_k^n - \Delta t \, g \left(\eta_{k+1}^n - \eta_k^n\right) \Delta x \tag{5.42}$$

$$\eta_k^{n+1} = \eta_k^n - \Delta t \, h_o \left(u_k^* - u_{k-1}^*\right) \Delta x \tag{5.43}$$

$$u_k^{n+1} = u_k^* \tag{5.44}$$

where n is the time level, Δt is the numerical time step, and Δx is the grid spacing. We assume that the computational domain covers the grid cells from $k = 1$ to $k = nx$ and that the cells $k = 0$ and $k = nx + 1$ are reserved for the implementation of boundary conditions. Care has to be taken here, given that the above equations are not symmetric with respect to the boundary conditions. The prediction for η does not use data of u_{nx+1} and the prediction of u does not use values of η_0. If we want to prescribe boundary conditions for η but not for u, this implies that u also needs to be predicted in the grid cell $k = 0$, which is assumed in the following.

5.13.5 Zero-Gradient Conditions

Zero-gradient conditions, also called von Neumann conditions, are sometimes employed for dynamic pressure at open boundaries for elimination for geostrophic flow components parallel to a boundary. For the barotropic surface gravity wave mode, being embedded in the dynamics, this condition implies vanishing flow normal to the boundary ($u_{nx} = 0$ and $u_0 = 0$). Hence, these conditions imply full wave reflection at lateral boundaries resulting is a standing wave that can significantly bias the predictions in the interior of the model domain.

5.13.6 Radiation Conditions

How can we achieve that a gravity wave escapes through a lateral boundary with no or only little reflection? The analytical solution of Eqs. (5.40) and (5.41) for a single wave is given by:

$$\eta = \eta_o \sin\left[\frac{2\pi}{\lambda}(x - ct)\right] \tag{5.45}$$

$$u = \sqrt{\frac{g}{h_o}}\eta_o \sin\left[\frac{2\pi}{\lambda}(x - ct)\right] \tag{5.46}$$

$$c = \pm\sqrt{gh_o} \tag{5.47}$$

where η_o and λ are amplitude and wavelength of the disturbance, and $c = \lambda/T$ (T is wave period) is the propagation speed of the disturbance. These solutions imply that a single sinusoidal wave pattern moves at a speed of c toward a lateral boundary.

On the other hand, movement of a signal ψ at speed c can be described by the advection equation:

$$\frac{\partial \psi}{\partial t} = -c\frac{\partial \psi}{\partial x} \tag{5.48}$$

For a single wave of known phase speed c, this equation can be used to predict changes of the boundary value via use of an advection scheme. In this context, Eq. (5.48) is known as the Sommerfeld radiation condition (Sommerfeld, 1949). With use of the upstream scheme, for instance, boundary data can be predicted with:

$$\psi_b^{n+1} = \psi_b^n(1 - C) + C\psi_{b-1}^n \tag{5.49}$$

where b refers to the boundary grid cell, $b - 1$ is the adjacent grid cell, and $C = |c|\,\Delta t \Delta x$ is the Courant number. This equation is replaced by a zero-gradient condition if the direction of c is directed into the model domain. Obviously, $C \le 1$ is required for the sake of numerical stability. The propagation speed c, however, is generally not known. For more complex processes, this speed can be a complex superposition of various wave types (e.g. gravity waves, Kelvin waves and Rossby waves) appearing as barotropic and/or baroclinic wave modes. To overcome this problem, the propagation speed in Eq. (5.48) can be estimated using the relation:

$$c = -\frac{\partial \psi/\partial t}{\partial \psi/\partial x} \tag{5.50}$$

This approach, first proposed by Orlanski (1976), requires reference to interior grid-point values at preceding times. An example is:

$$C = \frac{\psi_{b-1}^{n} - \psi_{b-1}^{n-1}}{\psi_{b-2}^{n-1} - \psi_{b-1}^{n-1}} \tag{5.51}$$

This expression can be substituted into Eq. (5.49) yielding the boundary condition:

$$\psi_b^{n+1} = \frac{\psi_b^{n}\left(\psi_{b-1}^{n} - \psi_{b-2}^{n-1}\right) - \psi_{b-1}^{n}\left(\psi_{b-1}^{n} - \psi_{b-1}^{n-1}\right)}{\psi_{b-1}^{n-1} - \psi_{b-2}^{n-1}} \tag{5.52}$$

It is worth mentioning that Orlanski's original approach led to the scheme:

$$\psi_b^{n+1} = \frac{\psi_b^{n-1}\left(\psi_{b-1}^{n} - \psi_{b-2}^{n-1}\right) - \psi_{b-1}^{n}\left(\psi_{b-1}^{n} - \psi_{b-1}^{n-2}\right)}{\psi_{b-1}^{n-2} - \psi_{b-2}^{n-1}} \tag{5.53}$$

Miller and Thorpe (1981) describe other schemes approximating Eq. (5.50). The implementation of radiation conditions implies use of more than two time levels. As outlined above, radiation conditions should be applied to dynamic pressure only, whereas velocity components normal to open boundaries should be directly predicted with the numerical scheme. Three-dimensional applications including the Coriolis force require additional conditions for flow running parallel to an open boundary.

5.13.7 Sponge Layers and Low-Pass Grid Filters

Various methods can be used to eliminate unwanted dynamic disturbances in close vicinity of a downstream boundary. Gradual increase of bottom friction towards an upward boundary is a common approach. In case of quasi-geostrophic flows, however, this can lead to unwanted flow convergence/divergence in the bottom Ekman layer. The associated modification of the surface pressure field can modify the geostrophic flow field in the entire model domain. Hence, bottom-friction enhancement should not be employed in geostrophic applications. A better option is to employ Rayleigh damping (see Eq. 5.39) to selected velocity components using a damping parameter that increases from zero at some distance from the boundary to a maximum value at the boundary. Finding the "best" value is often a time-consuming trial-and-error task.

Low-pass grid filters, on the other hand, are based on the fact that a numerical grid can only resolve disturbances of a wavelength exceeding the grid spacing. Gradual increase of the grid spacing towards an open boundary therefore gradually filters away wave disturbances. When using this method, the rule of thumb is that grid spacings should not increase by more than 10% from one grid cell to the next to avoid numerical problems. It is obvious that low-pass grid filters should only be used near outflow boundaries.

5.14 Final Remark

All fluid processes described in this book are governed by different versions of the *Navier-Stokes equations*. The numerical model, gradually developed from Exercise 1 through to Exercise 25, is backward compatible, which implies that the same model applied to simulate equatorial Kelvin waves or the abyssal circulation can be used to simulate short gravity waves or small-scale convective plumes. That the Navier-Stokes equations, which can be written down on a sheet of paper in a few lines, embrace such a rich variety of hydrodynamic processes is truly fascinating.

5.15 Technical Information

This book has been written in LaTeX using TeXnicCenter that can be downloaded from:

http://www.toolscenter.org/

in conjunction with MikTeX (Version 2.5) – a very nicely prepared LaTeX implementation for Windows platforms which can be downloaded at:

http://miktex.org/

Most graphs of this book were created with SciLab. Most sketches were made with Microsoft Word. GIMP has been used for the manipulation of images. GIMP is a cost-free alternative to commercial graphical manipulation programs such as Adobe Photoshop Elements. This software can be freely downloaded from:

http://www.gimp.org/

Bibliography

Arnold, J. R., and W. F. Libby, 1949: Age determinations by radiocarbon content: checks with samples of known age. Science 110, 678–680.

Arakawa, A., and V. Lamb, 1977: Computational design of the basic dynamical processes of the ucla general circulation model. In Methods in Computational Physics, vol. 17. Academic Press, San Francisco, pp. 174–267.

Bénard, H., 1900: Les tourbillons cellulairs dans une nappe liquide. Rev. Gén. Sci. Pure Appl., 11, 1309–1328.

Bernoulli, D., 1738: Hydrodynamica, J. R. Dulsecker, Strasbourg.

Bourlès, B., et al. 2003: The deep currents in the Eastern Equatorial Atlantic Ocean. Geophys. Res. Lett., 30(5), 8002, doi:10.1029/2002GL015095.

Britter, R. A., and P. F. Linden, 1980: The motion of the front of a gravity current travelling down an incline. J. Fluid Mech., 99(3), 531–543.

Broecker, W. S., and T.-H. Peng, 1982: Tracers in the Sea, Eldigio Press, Palisades, NY, 690pp.

Broecker, W. S., 1999: What if the conveyor were to shut down? Reflections on a possible outcome of the great global experiment. GSA Today, 9(1), 1–7.

Brown, 1866: Miscellaneous Botanical Works Vol. 1 (edited by J. J. Bennett), R. Hardwicke, London

Chapman, D. C., 2002: Deceleration of a finite-width, stratified current over a sloping bottom: frictional spindown or buoyancy shutdown? J. Phys. Oceanogr., 32, 336–352.

Courant, R., K. Friedrichs, and H. Lewy, 1928: Über die partiellen Differenzengleichungen der mathematischen Physik. Mathematische Annalen, 100 (1), 32–74.

Cushman-Roisin, B., 1994: Introduction to Geophysical Fluid Dynamics, Prentice Hall, New York, 320pp.

Deleersnijder, E., J.-M. Campin, and E. J. M. Delhez, 2001: The concept of age in marine modelling. 1. Theory and preliminary model results. J. Mar. Syst., 28, 229–267.

Dengler, M., F. A. Schott, C. Eden, P. Brandt, J. Fischer, and R. J. Zantopp, 2004: Break-up of the Atlantic deep western boundary current into eddies at 8° S. Nature, 432, 1018–1020.

Ekman, V. W., 1905: On the influence of the Earth's rotation on ocean-currents. Arkiv for Matematik, Astronomi och Fysik 2(11), 1–52.

Euler, L., 1755: Principes generaux de l'etat d'equilibre des fluides; Principes generaux du mouvement des fluides; Continuation des recherches sur la theorie du mouvement des fluides. Histoire de l'Acacdernie de Berlin.

Flinders, M., 1814: A voyage to Terra Australis. 1.

Fringer, O. B., S. W. Armfield, and R. L. Street, 2005: Reducing numerical diffusion in interfacial gravity wave simulations. Int. J. Numer. Meth. Fluids, 49(3), 301–329, doi: 10.1002/fld.992.

Froude, W.: 1874: On useful displacement as limited by weight of structure and of propulsive power. Trans. Inst. Naval Architects, 15, 148–155.

Gill, A. E., 1982: Atmosphere-Ocean Dynamics, Academic Press, London, 662pp.

J. Kämpf, *Advanced Ocean Modelling*, DOI 10.1007/978-3-642-10610-1,

Helmholtz, H. L. F. von, 1868: Über discontinuierliche Flüssigkeits-Bewegungen [On the discontinuous movements of fluids], Monatsberichte der Königlichen Preussichen Akademie der Wissenschaften zu Berlin [Monthly Reports of the Royal Prussian Academy of Philosophy in Berlin], vol. 23, pp. 215ff.
Kochergin, V. P., 1987: Three-dimensional prognostic models. In: Heaps, N. S. (ed.), Three-Dimensional Coastal Ocean Models, Coastal Estuarine Science Series, vol. 4, Union, American Geophysical Washington, DC, pp. 201–208.
Kämpf, J., 2009: Ocean Modelling for Beginners (using open-source software), Springer, 175pp.
Kowalik, Z., and T. S. Murty, 1993: Numerical Modelling of Ocean Dynamics, World Scientific, London, 481pp.
Kundu, P. K., 1990: Fluid Mechanics, Academic Press, Orlando, Florida, 638pp.
Laplace, P.-S., 1775: Recherches sur plusiers points du Systéme du Monde. Mém. Acad. Roy. des Sciences, 88, 75–182.
Marshall, J., and F. Schott, 1999: Ocean-ocean convection: observations, theory, and models. Rev. Geophys., 37(1), 1–64.
Mann, K. H., and J. R. N. Lazier, 1996: Dynamics of Marine Ecosystems: Biological-Physical Interactions in the Oceans, 2nd edition, Blackwell Science, Oxford, 394pp.
Maier-Reimer, E., 1980: On the formation of salt water wedges in estuaries. Lecture Notes on Coastal and Estuarine Studies, 1, 91–101.
Maier-Reimer, E., and J. Sündermann, 1982: On Tracer Methods in Computational Hydrodynamics. Engineering Application of Computational Hydraulics, 1, Pitman Advanced Publishing Program, Boston.
Miller, M. J., and A. J. Thorpe, 1981: Radiation conditions for the lateral boundaries of limit-area numerical models. Quart. J. R. Met. Soc., 107, 615–628.
Ohshima, K. I., and M. Wakatsuchi, 1990: A numerical study of barotropic instability associated with the Soya Warm Current in the Sea of Okhotsk. J. Phys. Oceanogr., 20, 570–584.
Orlanski, I., 1976: A simple boundary condition for unbounded hyperbolic flows. J. Comput. Phys., 21, 251–259.
Pacanowksi, R., and S. G. H. Philander, 1981: Parameterization of vertical mixing in numerical models of tropical oceans. J. Phys. Oceanogr., 11(11), 1443–1451.
Poisson, S. D., 1813: Remarques sur une équation qui se présente dans la théorie des attractions des sphéroides" Nouveau Bull. Soc. Philomathique de Paris , 3, pp. 388–392.
Pond, S. G., and L. Pickard, 1983: Introductory Dynamical Oceanography, 2nd edition, Pergamon Press, Oxford, 329pp.
Prandtl, L., 1925: Bericht über Untersuchungen zur ausgebildeten Turbulenz. Z. Angew. Math. Mech., 5, 136–139.
Press W. H., B. P. Flannery, S. A. Teukolsky, and W. T. Vetterling, 1989: Numerical Recipes. The Art of Scientific Computing. Cambridge University Press, Cambridge, 702pp.
Rayleigh, L., 1916: On convection currents in a horizontal layer of fluid when the higher temperature is on the other side. Philos. Mag., 32, 529–543.
Richardson, L. F., 1920: The supply of energy from and to atmospheric eddies. Pros. Roy. Soc. London, A 97, 354–373.
Ruddick, B. 1983: A practical indicator of the stability of the water column to double-diffusive activity. Deep-Sea Res., 30, 1105–1107.
Smagorinsky J., 1963: General circulation experiments with the primitive equations – I. The basic experiment. Mon. Weather Rev., 91, 99–165.
Sommerfeld, A., 1949: Partial differential equations. Lectures on Theoretical Physics, vol. 6. Academic Press, San Diego, CA.
Stommel, H., 1958: The abyssal circulation. Deep-Sea Res., 5, 80–82.
Stommel, H., and A. B. Arons, 1960: On the abyssal circulation of the world ocean – I. Stationary planetary flow patterns on a sphere. Deep-Sea Res., 6, 140–154.
Stommel, H. M., 1987: A View of the Sea. Princeton University Press, Princeton, N J, 165pp.

Stommel, H. M., and D. W. Moore, 1989: An Introduction to the Coriolis force, Columbia University Press, New York, 297pp.
Thomson, W. (Lord Kelvin), 1871: Hydrokinetic solutions and observations. Phil. Mag., 42, 362–377.
Tomczak, M., and J. S. Godfrey, 2003: Regional Oceanography: An Introduction, 2nd edition, Daya Publishing House, Delhi, 390pp.
Turner, J. S., 1973: Buoyancy Effects in Fluids, Cambridge University Press, Cambridge, 367pp.
Wakatsuchi, M., and K. I. Ohshima, 1990: Observations of ice-ocean eddy street off the Hokkaido coast in the Sea of Okhotsk through radar images. J. Phys. Oceanogr., 20, 585–594.
Wu, W., 2007: Computational River Dynamics, CRC Press, USA, 508pp.

List of Exercises

Index

MIX
Papier aus verantwortungsvollen Quellen
Paper from responsible sources
FSC® C105338
FSC
www.fsc.org

If you have any concerns about our products,
you can contact us on
ProductSafety@springernature.com

In case Publisher is established outside the EU,
the EU authorized representative is:
Springer Nature Customer Service Center GmbH
Europaplatz 3, 69115 Heidelberg, Germany

Printed by Libri Plureos GmbH
in Hamburg, Germany